# GARDENING
## à la Mode

# VEGETABLES

# GARDENING

## à la Mode

# VEGETABLES

### HARRIET ANNE DE SALIS

DOVER PUBLICATIONS, INC.
MINEOLA, NEW YORK

*Bibliographical Note*

This Dover edition, first published in 2017, is an unabridged republication of the work originally published by Longmans, Green, and Co., London and New York, in 1895.

*Library of Congress Cataloging-in-Publication Data*

Names: De Salis, Mrs. (Harriet Anne)
Title: Gardening à la mode : vegetables / Harriet Anne De Salis.
Other titles: Vegetables
Description: Dover edition. | Mineola, New York : Dover Publications, Inc., 2017.
Identifiers: LCCN 2016049208| ISBN 9780486814940 | ISBN 0486814947
Subjects: LCSH: Vegetable gardening—England. | Vegetables. | Cooking (Vegetables)
Classification: LCC SB322 .D4 2017 | DDC 635.0942—dc23 LC record available at https://lccn.loc.gov/2016049208

Manufactured in the United States by LSC Communications
81494701    2017
www.doverpublications.com

# PREFACE

———◦◦◦———

THESE little manuals on gardening are simply
intended to help amateurs; they do not pretend to
go deeply into the science of Vegetable and Fruit
cultivation, as there are so many standard works
on this interesting subject. These are merely
proposed to be handy little books of reference for
those persons who do their own gardening in a
small way, and are the upshot of the well-known
saying 'Experientia docet,' as when we came to
live in the country we were such Cockneys, we
knew absolutely nothing of gardening; and, as we
had to make our garden, which was only a field,
and could not afford an experienced gardener, we
set to work to learn the art. We bought various
books, and took in weekly periodicals on the subject,
and experimented on the advice therein contained
until we found out for ourselves what succeeded
best; and I am proud to say our experiments have

been crowned with success : these little volumes are the result of the advice we followed. Of course in such abridged works I cannot write all I could wish ; still I hope they will prove of some use to my readers. But I recommend those who do their own gardening to purchase Johnson's ' Dictionary of Gardening ' as a reference ; also Messrs. Sutton's work on the Culture of Vegetables, and Watts' ' Orchard and Fruit Garden ' ; and to take in the following weeklies—' The Gardener,' ' Home and Farm,' ' Field, Farm, and Fireside,' each costing one penny.

H. A. DE SALIS.

HAMPTON LEA, SUTTON.

# GARDENING—VEGETABLES

## ARTICHOKES (GLOBE)

THE artichoke, originally a native of Italy, was introduced into England during the reign of Henry VIII., and it is generally believed that the name refers to the part which is not eaten, and which is called 'the choke,' though that is quite a fallacy, as it is merely the English way of spelling its French name, *artichaut*, which is explained by old writers, who say it is a corruption of its Arabic name (*alcocalos*), from its heads being shaped like a pine-apple.

Artichokes are strong growing plants, and will grow almost in any kind of soil as long as it is properly prepared ; but, though they prefer a good rich garden soil, the ground should be manured and trenched in the middle of January, and the land should be broken up deeply and left with a rough surface till the end of March, then forked over, and the suckers planted (suckers are better than seeds) early in April four feet apart. The suckers are best when about nine inches long. Good manure should be placed between the rows every autumn, and the plants covered with straw in severe weather in the winter.

To prepare them for planting, the brown hard part by which they are attached to the parent stem must be removed, and the large outside leaves should be taken off so that the heart appears above them. If the weather is favourable, it is a good plan to put them into a pan of water for a few hours before they are planted, especially if they have been separated from the parent stem for some time.

A large flower-pot should be placed over each, and they should be watered liberally every evening till the flower heads appear. About June all side shoots should be cut off, and all the care they will require during the summer is the frequent use of the hoe and occasionally a mulching of liquid manure. The plants will produce a succession of heads from July to October. In the autumn cut down the stalks which have produced heads, and place a thick covering of litter about the stems and roots to exclude frost.

One has to guard against the ravages of an insect called *Cassida viridis*, a very small beetle with a black and green body.

Globe artichokes are not adapted for small gardens, as they take up such an amount of room.

There are two varieties, the green globe and the purple globe. In the latter the heads are tinged with purple, and the scales curved inwards and compactly.

*RECIPES FOR COOKING*

## Artichokes à la Barigoule

Prepare and blanch four artichokes, take out the inside, squeeze the water out, and season them with a pinch of salt and one small pinch of pepper. Put six tablespoonfuls of oil in the frying-pan and fry the top of the leaves. Prepare one gill of 'fines herbes' as for sauce. Put into a quart stewpan four ounces of grated bacon, quarter of an ounce of butter, quarter of an ounce of flour, one gill of broth, and the gill of 'fines herbes.' Stir over the fire for five minutes; put a fourth part of this stuffing into each artichoke; place a thin slice of bacon two inches square on the top of each, tie round with string to keep them in shape; put them into a sauté pan with two gills of stock; bake in the oven for twenty minutes; ascertain if done; then dish up and serve.

## Artichokes à l'Italienne

These must be cut into quarters and boiled in water enough to enable them to swim with ease, with a little salt and butter. When done, drain them well and lay them all round the dish with the leaves outwards. Then take some Italian sauce, with which mix a small piece of butter, and pour the sauce over the part that is to be eaten, but not over the leaves.

## Artichokes à la Provençale

Choose some artichokes that are very tender, cut them into four quarters, pare them nicely, and rub them over with lemon, that they may preserve their white colour. Throw them into cold water, the quantity to be in proportion to the size of the dish in which they are to be served. Trim a stew-pan with a little olive oil or butter, salt, and pepper, then put the artichokes all round, the bottoms downward, and set the whole to bake in a moderately hot oven. When done, drain the artichokes and serve them up with French melted butter on them, to which add a little glaze and the juice of a lemon ; or some sauce espagnole worked with a small lump of butter and the juice of a lemon.

## ARTICHOKES (JERUSALEM)

These are a tuberous-rooted variety, and are planted like potatoes in rows four feet apart in any soil, and succeed well in any odd part of the garden. The tubers will keep good in the ground during the winter, and may be dug up as required. Jerusalem artichokes are considered very nutritious, and certainly they are very delicious, and their flavour is very useful in seasoning many dishes.

The only attention Jerusalem artichokes require is an occasional hoeing to loosen the surface, and draw a little of the earth up round the stems. In August the stems should be cut off about the middle, so that they get more air and light.

They can be taken up in October, or as soon as their stems have withered entirely, and put into sand to preserve them for winter use.

In many situations the stems of these plants form a very useful screen during the summer and autumn.

### RECIPES FOR COOKING

### Jerusalem Artichokes à la Reine

Wash and wipe the artichokes, cut off one end of each quite flat, and trim the other into a point ; boil them in milk and water, lift them out the instant they are done, place them upright in the dish in which they are to be served, and sauce them with a good béchamel, or with nearly half a pint of cream thickened with a dessertspoonful of flour mixed with one and a half ounces of butter and seasoned with a little mace and some salt. When cream cannot be procured, use new milk and increase the proportion of flour and butter.

## Jerusalem Artichokes Fried

Boil them in plenty of water for about twenty minutes. Beat two eggs, season two ounces of fine crumbs of bread with a grain of pepper, a quarter of a grain of cayenne, and a tablespoonful of Parmesan cheese ; dip the artichokes into the egg and strew them over with the crumbs ; fry in butter to a pale brown colour eight minutes, and serve uncovered.

## ASPARAGUS

Asparagus was originally a wild sea-coast plant, is a native of Great Britain, and formerly grew wild in many parts of England and Scotland, but is now to be found all over the world, and is grown more largely in France than in other countries, large quantities being raised among the vines. It was a very favourite vegetable with the ancient Romans.

Asparagus is grown from seed, and it does not do well in a heavy soil. The ground should be prepared in February, and left till March roughly exposed to the weather ; then, after the frost has worked upon it, it should have a good top dressing with sand, clay, burnt earth, leaf mould, soot, also old stable manure, and worked in well, and the bed left so that the surface may become dry and sweet. A very excellent thing for an asparagus bed is to get sea-sand and plenty of seaweed to dress it with. The seeds should be sewn thinly about the end of March.

To make an asparagus bed, a good dressing of manure should be first applied on the soil, and then deeply trenched to a depth of three feet, leaving all the best soil at the bottom of the trench, when it should be broken up with a fork and then well drained. Asparagus beds cannot be too much manured ; it takes three years to raise asparagus from seed, therefore it is advisable for those who cannot wait so long to procure plants of three or four years' growth, and even then the asparagus should not be attempted to be cut its first year. Where seed is planted it should be drilled shallowly in fine mould and planted ten inches apart ; and all that is necessary for the first year is to keep the seeds down, and the next year, about April, the plants must be thinned out to nine inches apart, and the seedlings can be used to make another bed if required.

For the first two years it should be allowed to run to stalk.

In *planting* asparagus the greatest care is necessary, as the roots are so very delicate, and should be carefully spread out and the crowns of the plants be left just visible above the surface.

A writer of great experience says that ' not one gardener in twenty properly manages asparagus beds, as the soil should not be touched with fork or hoe, for fear of bruising the plants.

' It is most necessary to keep all weeds down. Salt, if given liberally three times a year, will effect this, besides being a first-rate fertiliser. Proper knives are sold for cutting asparagus, as great care is required in cutting it, so as to prevent wounding the plants by cutting invisible heads. Cutting may commence as soon as the heads are high enough to cut, and after June the stems will go to seed till the autumn, when those which are ripe should be cut down and the berries sprinkled over the beds and the stems cut off.'

The beds must be then arranged for the winter by dressing them with salt and good short stable manure, and over that some good rich soil, and flattened evenly with the back of the spade, and in the spring should have another dressing of salt, and the soil and manure left on.

The Battersea and Connover's Colossal are the favourite kinds to sow or plant, but the Argenteuil early purple and late purple are becoming very popular, and are cultivated in Argenteuil expressly for the French and English markets.

### RECIPES FOR COOKING

To cook asparagus it should be arranged in bundles, the heads together, and the tough part of the stalk should be removed before cooking. The stalks should be carefully washed, then tied together and put into boiling salted water deep enough to cover them. There is now what is termed an asparagus kettle, which is really a necessity to cook asparagus properly. The bundle is laid on a drainer, which fits into the kettle, and the cook is enabled to lift the cooked asparagus out of the water without breaking the heads. One or two tablespoonfuls of vinegar are sometimes put in the water : it helps to keep the green colour of the vegetable. The time for cooking it varies according to its age and freshness ; from ten minutes to half an hour is about the usual time. Where there is no proper asparagus kettle the asparagus should be stood up in the saucepan, so that the green tops will be at least an inch above the surface, that the stalks may be well cooked before the heads are broken.

When cooked they should be thoroughly drained and placed on a silver or china drainer in the dish in which they are to be served. There are now sold asparagus dishes, a kind of rack in electro-ware, which are most handy as well as ornamental.

## Asparagus and Eggs

Cut cold boiled asparagus into small pieces, put them into a buttered baking dish, season well, and drop eggs over the top without beating, and bake till the eggs are cooked.

## Asparagus (Italian Fashion)

Take some asparagus, break them into pieces, then boil them soft, and drain the water off ; take a little oil, water, and vinegar, let it boil, season it with pepper and salt, throw in the asparagus, and thicken with the white of eggs. Endive done this way is good ; green peas done as above are very good, only add a lettuce cut small and two or three onions, and leave out the eggs.

## Belgian Asparagus

Boil in the usual way as much asparagus as required, and arrange it neatly on a folded napkin in a flat dish. Boil some eggs hard (allowing one egg to each person), and divide them into halves lengthwise. Border the asparagus with these, placing them with the yolks upwards. Serve this dish very hot, and send to table with it a sufficient quantity of butter, simply melted and made quite hot, but without any thickening.

# AUBERGINES

This vegetable is not yet common in England, but it is becoming more generally appreciated, as the fruit makes many delicious dishes. Aubergines are the fruits of the egg-plant. Seeds may be sown in heat, in the same manner as vegetable marrows, and planted out in a sunny spot. In the summer the soil must be rich. There are two kinds, but, for table use, Melongena, which is the purple, is the one adapted for table culture. In every respect treat as for marrows. Such appetising dishes are produced from this vegetable that it is surprising it is not more grown in this country, and if there be space enough, I recommend all amateurs to try it.

*RECIPES FOR COOKING*

### Aubergines Farcies

Cut the aubergines down the middle lengthwise, scoop out the inside, taking care not to break the skin. What has been scooped out put into a stewpan with a couple of ounces of butter ; let it simmer till soft ; mash it up with a few breadcrumbs, a couple of hard-boiled eggs run through a wire sieve, a little pounded chicken, with pepper and salt to taste. Mix thoroughly, and stuff the aubergines with it, and brush over with beaten egg ; strew some crumbs on the top, and put a few dabs of butter on each. Bake in the oven a few minutes, till they become a nice brown colour ; then serve, and garnish with tufts of parsley. (Mrs. Grace Johnson.)

### Aubergines (Another Way)

Put the aubergines into the oven to roast, but they must not burn ; when soft, take them out, and cut them through ; likewise take out the inside carefully, mash it with butter, pepper, and salt ; fill in the skins again, and serve on a piece of fried bread. Garnish the tops with minced hard-boiled eggs and tufts of parsley. (Mrs. Grace Johnson.)

## Aubergines à la Parmesan

Peel the aubergines, cut them up into slices, sprinkle them over with salt, drain, and fry them quickly until *nearly* done, season, and cover them with white sauce ; then put the slices in layers in a white china fire-proof dish, and sprinkle each layer with Parmesan cheese ; baste with melted butter and put the dish into a moderate oven for a quarter of an hour.　(Dubois.)

## Aubergines à la Turque

Cut some aubergines into slices about half an inch thick, sprinkle them with salt, and let them remain about an hour. Put a couple of ounces of flour into a basin, and mix it to a smooth paste with a little water, but not too thick.　Then dip the pieces of aubergine into it, and fry them in butter till a nice golden colour ; when done, take them out, drain, and dish up.

## BALM

This herb thrives in any ordinary garden soil, and is propagated by division in the spring.

## BASIL

There are two kinds—the common, or sweet, and the bush basil.　March is the best time for sowing in a slight hot bed, and in May it can be planted out in a warm border in light soil.

## BEANS (BROAD OR WINDSOR)

The bean is said to be a native of Egypt, and is supposed to have been brought to England by the Romans.

The priests of Egypt held it a crime even to look at beans ; the very sight of them was unclean. Pythagoras forbade his disciples to eat beans because they were formed of the rotten ooze out of which man was created. The Romans ate beans at funerals with awe, from the idea that the souls of the dead were in them.

These beans are extremely nutritious, and, when gathered young and nicely cooked, are very digestible.

They succeed best in a deep, stiff, loamy, moderately rich soil, and, once the seed is sown, require little attention beyond earthing up the plants well by drawing the soil freely against them on either side when the young plants are a few inches high. They should be planted in single rows, with a distance of six to twelve inches from plant to plant. The dwarfs should have less space between them than the tall varieties.

The Long-pod varieties should be sown in February and March for early crops.

Immediately the plants have ceased blooming pinch off the points beyond the blooms, and if the weather is very dry, damp the blooms over with water from a syringe. Broad beans are subject to the attacks of the black fly. The best preventive of this pest is early planting and pinching out the points of the plant.

Norfolk Giant, Giant Seville Long Pod, and Green Windsor are some of the best sorts.

## Broad Beans au Jus

Take a sufficient quantity of young beans ready shelled, and put them on to boil with a great deal of salt and water ; when nearly done drain them, and put them into a stewpan with a piece of butter, some minced parsley, and chives ; toss them well in the butter ; add three spoonfuls of espagnole sauce, and let them simmer ; then skim off the fat, reduce the sauce, and serve.

## BUTTER BEANS

Butter, or wax-pod, beans hail from America, but are seldom found here. Yet they afford a nice variety, and it is well to give them a trial. Their specialty is that the pods are of a lovely golden colour with a semitransparent appearance, and, if gathered young, are perfectly stringless, and should be cooked whole without being sliced at all. The cultivation of this bean is exactly the same as that of the French and Runner Beans. The best sorts are the Dwarf Golden Butter, the climbing Mont d'Or, and the dwarf German Black Wax.

## FRENCH OR KIDNEY BEANS

The dwarf kidney bean was originally a native of India, and was introduced before the time of Gerard.

These beans require deep rich soil and a good amount of moisture with a warm sunny aspect. Sow in the open ground in May, as it is not safe to grow them before unless the seeds have been sown in frames beforehand. The seeds should be sown thinly or in double rows about four inches apart. The best course is to allow a few extra seeds for thinning, especially for the earliest crop.

The great thing in procuring really prolific and healthy French beans is not to grudge manure and to water with soot-water occasionally. Red spiders are very fond of these plants, and their ravages cannot be stayed if the plants are starved. The pods will turn yellow and drop off.

Should many gross leaves form and young shoots push through them, let both be neatly pinched off so as to admit more light to the roots and assure more pods and less foliage. Pick off all beans as soon as they become large enough to use.

Canadian Wonder is one of the finest kinds ; Ne Plus Ultra and Negro Long Pod are also very good, though, of course, there are several other good varieties ; but I have found the three quoted the best.

## RECIPES FOR COOKING

### French Beans à la Parisienne

Remove all fibres from some French beans, break off the ends, boil them in boiling water, then toss them in melted butter, seasoned with chopped chives and parsnips. Stir in a dessertspoonful of flour, a pinch of salt, and a gill of white stock. Reduce the sauce, thicken with two yolks of eggs, and flavour with a few drops of lemon juice just before serving.

### French Beans à la Poulette

Pick one pound of French beans, put them into a gallon stewpan with three quarts of boiling water and one pinch of salt, and boil till tender. Put into a two-quart stewpan one ounce of butter and half an ounce of flour, stir over the fire for three minutes, then add three gills of water and one pinch of salt ; boil for ten minutes, thicken with the yolks of two eggs and half an ounce of butter, drain the beans ; put them into the sauce with half a tablespoonful of chopped parsley ; mix, and serve.

## RUNNER BEANS (SCARLET RUNNERS)

The scarlet runner is a native of South America, and was not introduced till 1633, when it was first cultivated in the flower gardens only as an ornamental plant.

These beans, like the French, require a good rich soil, but the soil should be manured the previous autumn, as, if freshly manured, the plants will go mostly into wood and leaves. The seeds should be sown in a sunny situation about the end of May in a soil that has been trenched three feet in depth. Scarlet runners never succeed in a poor soil. Sowings made somewhat more thickly than ordinary are advisable, as it forwards and increases the crop.

The stakes or poles should be placed into the ground at a distance of two feet from each other before the seeds are put in, and around each pole plant six beans two inches deep. When about a foot high the plants must be well earthed up and tied loosely round the pole, winding them, in their natural way of climbing, from left to right. A very pretty way of growing them, where there is space enough, is to put the stakes down each side of a path, and when they have grown up they make a shady archway, and look very ornamental with white and red blossoms.

In very dry weather the roots ought to have an occasional soaking with water or liquid manure, and if, from the same cause, the pods fail to set, a good overhead syringing night and morning will generally enable them to do so.

Runners may be grown without sticks by pinching out the points of the shoots several times beyond the third or fourth leaf.

The best sorts to grow are Painted Lady and Sutton's Champion and the Czar ; Daniel's Giant is also very fine.

Scarlet runners may be cooked in any way French beans are.

## BEETROOT

This was a native of the sea-coast on the south of Europe, and was introduced into England in 1656. This vegetable likes the same rich, deep soil, and open and sunny situation, as the carrot. Soot of any kind or burnt material, salt, and nitrate of soda are the best manures for it, and should be applied to the soil before sowing, or as a top dressing after the plants are up. The ground should be dug two spades deep, and the whole of the manure intended for it should be put in with the bottom spit, so that it may be buried twelve inches within the ground. An early sowing should be made in April and a main sowing the first week in May. It is most necessary that the seed should not be put in till the severe frosts are over. The seed is best sown in drills a foot apart and an inch deep.

The seedlings should be thinned out about six or eight inches between the plants *at least*. The hoe should be occasionally used to give them air as well as to destroy any weeds. October is the time for taking it up for use as it is wanted, but it is best to wait till November or the beginning of December to take it up entirely, when it should be buried in sand or sifted coal ashes in alternate layers under shelter, after removing the leaves and fibrous roots. *Great* care must be exercised in taking it up not to break the roots, as, if any of the very smallest fibres are injured, the roots bleed, and both colour and quality suffer, as it will become a dingy whitish pink instead of a brilliant dark red.

Some of the best kinds are Dell's Crimson and Nutting's Dwarf Red. The former is very compact in growth and delicate in quality. There is a new variety called Beet Cheltenham Green Top, which has been very highly spoken of. It has green foliage, and is smaller than most other kinds. The root is a very deep shade of purple black, and the flavour very delicious.

*RECIPE FOR COOKING*

### Beetroot à la Crème

Peel a beetroot and cut it into slices, then cook it very slowly in a melted butter sauce or white béchamel. Season with pepper and salt, and serve.

## BORAGE

This is a very robust herb, though it prefers a light warm soil, but not rich, and an open situation. It is best sown in March and two or three times successively during the season, in shallow drills, about twelve inches apart, and, when about six weeks' growth, thinned out to eight or nine inches apart.

To gather the seed : After the flowering is over and the seed is perfectly ripe the stalks should be gathered and dried thoroughly, when the seeds can be rubbed out. Borage seeds will fly all over the place, and often in the next year clumps of borage will appear in different parts of the garden.

Borage is principally cultivated for the flavouring of claret and other cups.

## BORECOLE (*see* SCOTCH KALE)

## BROCCOLI

This vegetable is said to have emanated from Italy. There are now ten or twelve distinct kinds of broccoli at least, but all have sprung from the two kinds originally brought over—the purple and the green.

The *soil* should be very rich and the ground deeply trenched and manured.

The first sowing should be made early in March in a gentle heat, and if there be space, other sowings about once a fortnight, beginning the second week in April. The seedlings should be transplanted thickly at distances of about five inches apart as soon as they become large enough. They must be transplanted into drill rows at distances of three feet apart, and a similar distance allowed between each plant in the row.

They require good waterings during all dry weather. Frost is a great enemy to this vegetable, and, if severe, will injure it before any usable hearts are produced : these are always in danger when the thermometer falls to 10 degrees, except when sheltered by snow ; therefore the late autumn kinds should always be protected by means of bracken, fern, straw, &c., at the approach of frost.

Immediately the 'flower' is seen through the leaves the heads for use must be cut immediately. Sprouting broccoli should have the chief head cut off as soon as it is seen to have produced the necessary bulk.   The best kinds are Midwinter Snows and Winter White for autumn use, Leamington for early spring, and Sutton's Dwarf White, followed with Carter's Summer Late Queen and Sutton's Bouquet.

There are two kinds of sprouting broccoli— purple and white.   The purple is the hardier and far more productive, and if sown at periods from April to the end of June will be in use from November to April ; the flavour is unequalled. North's early purple is one of the best kinds.

## BRUSSELS SPROUTS

These are one of the most useful of winter vegetables, and should be sown in frames in February or in shallow boxes placed on a greenhouse shelf, and for succession in April in the open ground, then early in March the plants should be pricked out.   The *soil* should be rich and light, and requires liberal manuring.

The seeds are best procured direct from Brussels.   They should be planted one foot apart every way.   They are never good until there has been some frost upon them, and therefore should not be cut till then. The leaves should not be cut before the sprouts are formed, but the crowns should be pinched out in September, so that the whole strength may be thrown into the sprouts.

The best kinds are Scrymger's Giant and Dalkeith.

Later sowings should be made in March, April or even as late as May for successional supplies, but where there are only small gardens it is best to buy the young plants and plant them out. The ground should be deeply worked and made *very firm*, and should not be very rich with fresh manure, or the growth will be too rapid and soft to endure severe weather. In rich loamy soils the plants should be placed two feet apart in rows from two feet to two and a half feet, but in shallow or poor ground which is only worked to the depth of a single spade, a spade of eighteen inches one way and two feet the other will be sufficient. Always begin cutting the sprouts from the bottom, and after all the sprouts are gathered, the heads may be cut off, as they form a very excellent dish. The best kinds are Scrymger's Giant, the Dalkeith, the Aigburth, also Wroxton, and Sutton's Exhibition.

## RECIPES FOR COOKING

### Brussels Sprouts au Jus

Boil them a few minutes in water, and then stew them till tender in some good gravy with a little salt and pepper. They may also be served with a white sauce. Boil them first, and then top them in a frying-pan with a little butter. They must not brown.

## Sprouts à la Française

Trim the sprouts, and place them into salted water, and cook them in a saucepan with the cover off. When cooked, drain them as dry as possible, and to each pound of them allow an ounce of melted butter, and when it is boiling lay it on the sprouts, and let them cook for five minutes on the side of the fire, shaking the pan constantly ; then serve as hot as possible.

## BURNET

is a hardy plant possessing both the odour and flavour of cucumbers. It can be sown in any light soil in March or April and then thinned out to six inches apart.

## CABBAGE

The cabbage in its wild state is biennial, and grows naturally on the sea coast in different parts of England, and, according to Mrs. Loudon, belongs to the same genus as broccoli, cauliflowers, Scotch or German greens, brussels sprouts and savoys, and not only to one genus, but are actual varieties of one of the species of the genus, viz., *Brassica oleracea*, and that the turnip and Swedish turnip belong to other species of the same genus.

The word ' cabbage' means a firm head or ball of leaves folded closely over each other.

All cabbages require a rich, well-worked soil, which should be constantly hoed after the seed is sown. The early varieties should be sown thinly in frames in March, and the seed just covered lightly over with the soil, and the frame kept dark and closed till the plants come up, when the frames can be opened in the daytime if the weather is favourable, and later on cautiously during the night. They should be well watered with tepid water when dry, and thinned out, so that they should not touch one another.

Always press the soil well down after planting. Seed can be sown in the open ground in April, and will be ready to thin out in the middle of June. Savoys can also be repeated in the middle of July to the middle of August.

When ready for planting, the smaller growing varieties should be set eighteen inches apart in the rows, with the same distance between, but the larger growing sorts must be planted two feet. They require frequent hoeing.

It is very necessary that all old stumps of cabbages should be cleared away and burned, and the ashes returned to the ground. This is a great preventive to clubbing, as great numbers of larvæ are thus destroyed.

Cabbages are often attacked by caterpillars of the common dart moth; they will hide under clods and in cracks in the earth, and by watering round the plants thoroughly with soapy water they will often show themselves. Soot scattered round the plants and chopped in with a hoe will keep these pests at bay. Cabbages are also liable to mildew and ambury, and many insects, such as aphis mamestra, cabbage fly and cabbage butterfly, cabbage moths, and cabbage garden pebble moths.

Ambury is also called club root, and appears in the shape of a wart on the stem close to the roots, which contains a white maggot, the larvæ of the weevil. The wart should be removed and the plant placed back in the earth. This pest mostly attacks the cabbage when grown for successive years on the same soil, often whole rows of cabbages are considerably damaged by them and slugs. Soot and lime lightly forked into the ground previous to peaching and placed round the roots is a great preventive of this disease, and in planting out it is also an excellent plan to dip the roots in a puddle made of soot, lime, and soil. All excrescences should be pinched before planting, as they invariably hold a maggot. The cabbage moth or mamestra haunts the gardens in May and June, in the evening time. The caterpillar of it is green, marked with a dark stripe down the back and a yellowish one down each side, and is to be found in July, August, and September feeding upon the hearts of cabbages.

The ' Gardeners' Chronicle ' recommends the handpicking of the caterpillars as the only cure.

The cabbage fly is the parent of a maggot which causes fearful havoc among them, and when the cabbages become yellow in colour, and droop at midday from the effects of the sun, the fly is doing its worst, and the plants should be taken up and burnt, and lime put into the holes. When cabbages are cut, the stem should be cut crosswise, and then innumerable young sprouts will form, which make very tender eating.

The best kinds to grow are :—

Early York—Good and small ; requires good cultivation and to be grown quickly.

Enfield Market—A *very good* cabbage.

Ellam's Early—The finest of the small early class, very dwarf.

Early Rainham—Early fine cabbage, rather large.

St. John's Day—Good.

Sutton's Imperial—Very good all round cabbage.

Sugarloaf—One of the best for late sowing.

Wenningstadt—Very large, produces firm hearts of a conical shape and superior quality.

Pomeranian—Tall and tender, producing very solid tapering heads.

Hippocrates used cabbage very much for medicinal purposes. When any of his patients was seized with colic he always prescribed a dish of boiled cabbage with salt. Erasistratus considered it as a sovereign remedy for paralysis, and Pythagoras wrote books on the marvellous virtues of the cabbage.

**Coleworts.**—Coleworts are young cabbages gathered before they form a head, and are generally sown in June or July for autumn, winter, or early spring.

As they are always eaten young it is not necessary to plant them more than ten inches apart every way; and when they are gathered the stalks are pulled up and thrown away. In every respect they should be treated the same as cabbages. Rosette Colewort is one of the best varieties.

**Savoy Cabbage.**—These are large cabbages with wrinkled leaves, and should be sown at the end of February, when the plants should be ready to prick out in April and for planting out at the end of May, and then another sowing in the middle of March, pricked out in May, and planted in June; these are for autumn and early winter supplies; for the main crop they must be sown at the end of April or early in May, and pricked and planted out after similar intervals when required for winter and spring use.

The culture is the same as that of cabbages, except that they should be planted two feet apart every way. To prevent clubbing, deep culture and frequent changing of crops are the greatest safeguard. Dip the roots of all young plants into a lye formed of cow-dung, wood, ashes, &c., just before planting them into their winter quarters. Lime and soot sprinkled amongst the seedling plants when very young often prevent pests from attacking them as it does when small. In the cultivation of *all* cabbages one has to guard against the depredations of the cabbage butterfly, which are especially destructive in the larva state. These caterpillars are soft, of a pale whitish green, and very active ; the chrysalis is green, and looks like a swathed-up mummy. The best sort of Savoys to grow are, for small gardens, Sutton's Tom Thumb, the Dwarf Green Curled, and Dwarf Ulm. Sutton's Golden Globe is also a very fine sort, with bright golden coloured hearts. Norwegian is a very good variety for cold climates and late use.

There are two other kinds of cabbage much liked, the Couve Tronchuda, which is a Portuguese cabbage, and the Chou de Burghley, which originated by crossing a cabbage with a broccoli. The seeds of these are sown from February to April, and require a rich soil with plenty of space, as they have very large leaves, the midribs of which are cooked in the same way as sea kale. These seeds are to be procured at Sutton's.

**Red Cabbage.**—The cultivation is the same as for other cabbages. The best sorts are Sutton's Dwarf Blood Red and New Early Red Nonpareil. They should not be gathered till quite hard, and have been touched by frost.

*RECIPES FOR COOKING*

## Cabbage Ragout

Take the half of a middle-sized cabbage, boil it for half an hour, and then change it in cold water ; squeeze it well, and take out the heart ; cut the cabbage into small pieces, and put it into a stewpan, with a slice of good butter, turn it a few times over the fire, and shake in some flour, put in sufficient gravy to give colour to the ragout, then let it boil over a slow fire until the cabbage is done and reduced to a thick sauce ; season it with salt, a little coarse pepper, a little grated nutmeg, and serve under any roast meat or stewed sausages.

## Chou à la Bourgeoise

Boil a whole cabbage, well cleaned, for a quarter of an hour ; lay it in cold water ; when cold, take it out and squeeze it dry ; open the leaves carefully, and between each put a little veal forcemeat ; tie all together with string, and stew it in as much broth as will cover the cabbage, with salt, pepper, sweet herbs, a bay leaf, an onion, carrot, and parsnip ; also add two or three cloves. When stewed enough, press it gently with a clean cloth, remove the string, cut it into halves, and serve with sauce espagnole, poured hot over it.

## Chou Rouge au Vin

Slice finely a red cabbage, blanch it in boiling water, and drain. Cut some bacon and chop it up small, and put it into a stewpan with some butter, and, when hot, place in the shredded cabbage with some good stock and some sauterne or chablis ; when quite tender it is done, and can be served.

## Red Cabbage à l'Allemagne

Wash, cut up, and drain a red cabbage ; put it into a stewpan with a good-sized piece of butter, and fry it slightly then add a cupful of good gravy, salt and pepper, a spoonful of good brown vinegar, and stew gently till tender.

# CAPSICUMS

These must be sown on a hot bed at the beginning of April, or can be raised on a warm boiler under handglasses, but then the sowing must not take place till towards the end of May. The young plants should be thinned to four inches apart, when they have still their seed-leaves, and plant in 4-inch pots, three in each, and keep in a moderate heat, shaded from the noonday sun and watered slightly with tepid water till they have taken root.

Air must be admitted freely to prevent their being drawn, and as May advances they must be gradually exposed to the sun by taking away the glasses in the daytime and leaving them by degrees open of an evening, then early in June they will be ready to plant out two feet apart, screened from the sun, and well watered till rooted, and the watering should be continued in dry weather. The blossom appears in July or early in August, when the pods will be ready to gather for pickling at the end of August. When in flower they are very ornamental and look well in a conservatory, and are grown for both use and ornament : some are red and some yellow.

Sutton's Erect-fruiting is one of the best, being

very ornamental, its colour a deep crimson. Red Cherry and Red Giant are also very fine. Of the yellow variety Golden Dawn and Prince of Wales are the best.

## To Pickle Capsicums

The capsicums should be put into a jar, and covered with boiling vinegar, allowing a good teaspoonful of salt and half an ounce of powdered mace to every quart of vinegar. When cold, tie down closely with a bladder. They will be ready to use in five or six weeks. (Cassell.)

# CARDOONS, or SPANISH ARTICHOKE

These are natives of Candia, and require a light, rich, moist, *sunny* soil, dug deep and well pulverised.

The seed should be sown at the end of April for an early crop, and a later one in June. Trenches should be made for it the same as for celery, and the seed arranged in groups of intervals of fifteen inches in the rows, and well supplied with water. When about a month old, the seedlings should be thinned out to four inches apart ; when advanced enough to be planted out, they must then be taken up carefully, and the long straggling leaves removed. When the plants are fully grown, which will be in August for the first sowing, and in October for the second, the leaves must be drawn together and surrounded with hay or straw bands to blanch them, and they must be earthed up like celery, which must be done in dry weather. It takes about ten weeks to blanch them. In severe weather litter should be thrown

over them to preserve them through the winter.
The stalks of the inner leaves are used for stews,
soups, and salads.   If these cardoons cannot be left
in the ground for want of space, they should be
placed in a cellar, stuck in beds of sand until the
stalks are bleached.

The spineless leaved Spanish cardoon is better
to grow than the French variety.


## RECIPES FOR COOKING

### Cardoons

are dressed in various ways.   Boil them until soft in salt and
water, dry them, butter them, and fry a good colour ; then
serve with melted butter.   They can be boiled and worked
up in a fricassee sauce, or they may be tied up and dressed
as asparagus.   To stew them, cut them into pieces, and stew
in white or brown gravy ; season with ketchup, salt, and
cayenne, and thicken with a small lump of butter rolled in
flour.


### Cardoon Salad

Boil some cardoons in an earthenware vessel till tender,
having stripped off all leaves except the white ones, and cut
them into lengths of three or four inches.   When tender,
yet firm, they must be strained and wiped dry, and put into
a stewpan, and sprinkled with pepper and salt.   In the
meantime, heat some oil in a pipkin with a piece of garlic,
and, when hot, add a quarter of a tumblerful of white wine
vinegar, well peppered ; let it boil, then pour it over the
cardoons, and leave over the fire for two minutes ; sprinkle
with chopped chives, and place into the salad bowl.

## CARROTS

Carrots were introduced into England in the reign of Elizabeth, and were so highly esteemed that the ladies wore leaves of it in their headdresses.

Carrots require deep, light, warm soil, well trenched but not very recently manured.

The seeds should be sown in dry weather, as if a shower of rain falls just after sowing it is almost fatal.

Sowing in the open ground should not commence before April, as the seeds will not germinate in a low temperature ; and if frames are available it is best to sow broadcast in them in January or February on a hot bed, and cover the glass with mats till the seed appears. If the soil gets dry, a slight watering should be given, but no damp must be allowed to enter. The plants should be thinned when they get four leaves, and must be placed half an inch apart, and air given them when the weather permits. The heat should not be more than 60° in the daytime and 50° at night.

For open air sowing, the seed should be sown in drills eight to twelve inches apart, according to the sorts, and as the carrots make their appearance, hoeing, weeding, watering, and thinning them to half an inch apart should be attended to, and when the plants are the size of a lead pencil, thin them to three or four inches apart, and before the last thinning is completed amongst them there should be small delicate carrots fit for the table.

When carrot leaves begin to change colour the roots should be taken up and the tops cut off, and the carrots placed in a cellar or outhouse and buried in sand.

There are two kinds of carrots, the long and the horn.

The best kinds to grow are : Sutton's Early Gem, Early French Nantes, Sutton's Champion Scarlet Horn, French Horn, and Telegraph.

## RECIPES FOR COOKING

### Carottes à l'Orléans

Take a few young carrots, trim them of an equal size, cut them into slices about the eighth of an inch thick, and blanch them well. Next lay them on a towel to drain ; put them into a stewpan with a small lump of sugar and a little broth, and let them boil over a large fire. When reduced to glaze add a good piece of fresh butter and a little salt. Care must be taken that the butter does not adhere to the carrots when served, as no sauce must be seen.

### Croquettes de Carottes

Take half a dozen carrots, boil them tender, and mash them whilst hot to a pulp, then add two yolks of eggs, salt, and white pepper, a grate of nutmeg, cinnamon, and moist sugar ; beat all well together, and make into the shape of croquettes ; dip into the white of eggs and fine breadcrumbs, then fry in boiling oil till of a pale golden colour.

## CAULIFLOWER

Soil required for cauliflowers should be very rich, with any amount of root moisture. The ground should be heavily manured, and with good management, a sheltered garden, frames, and hand-lights a complete succession can be had the whole year, excepting perhaps if there is a continuance of frost during the winter, especially in January, when it would be difficult to obtain supplies.

The earliest cauliflowers are obtained by sowing seed out of doors at the end of August, placing them into small pots and wintering them in a cold frame, and then planting them out on a sheltered border about the middle of March.

In the middle of February another crop should be sown in gentle heat, pricking them out in boxes of light soil in frames, and plant out the second week in April, and in the middle of April a third sowing should be made, and then there will be good heads in August and September, and the latter will continue the supply if dug up and laid in soil in pits or cold frames.

Two feet asunder should be allowed between the plants when planting out. Of course where space is limited, and there is not room for successions, the sowings must be less frequent.

The soil should be well watered at sowing time, and when the plants are pricked out, dry weather should be selected, and the time late in the afternoon. Frost is a great destroyer of cauliflowers, and it is well in November to pull up the plants and tie the leaves over their heads and bury them in sand, laying them in alternate layers with the earth in a dry situation. Some gardeners, after pulling them up and tying the leaves over their heads, merely hang them in a shed or cellar, which keeps them good for some time.

Spring sowings are liable to the inroads of White Fly, which cluster on them and eat their centres, and cause what is called 'blindness.' As a preventive, sprinkle soot over the leaves when damp, or the hearts with tobacco powder.

When a cauliflower is fit for use, which is known by the border opening as if about to run, *pull* up the plant at once. A cauliflower should be cut early in the morning, before the dew has gone, as if left to the middle of the day it loses much of its firmness and boils tough.

The best kinds to sow are, for August sowings : Early London. First Crop, or Magnum Bonum, for middle of February. For April sowing : Sutton's King of Cauliflowers (Dwarf), Sutton's Amateur's Pride, and Veitch's Autumn Giant. Walcheren is also an excellent mid-season cauliflower.

### RECIPE FOR COOKING

### Cauliflower à la Burmah

Cleanse a large white cauliflower, cut the stalks close to the bottom, leaving a wreath of the short stalks all round. Boil it in milk and water till it is tender. Make a white sauce, into which add some grated parmesan cheese and some curry powder ; bring to the boil, and then simmer till the sauce is smooth, when it can be poured over and round the cauliflower.

## CELERIAC, or TURNIP-ROOTED CELERY

This, where there is room for it, is a very useful vegetable, and is largely used abroad, but until quite lately has not been much cultivated in England. In flavour it resembles considerably the ordinary kinds of celery, the difference principally being that in the latter the blanched leaf stalks constitute the edible portion, and in the turnip-rooted variety the tuber alone is used, which is generally cut into slices and served in salad, though it is sometimes boiled and served with white sauce ; it is also very good for flavouring soups.

The seed should be sown in March, and in the same way as celery, though the culture of it is more simple. The drills should be six inches apart, and the ground well watered morning and evening in dry weather. Prick the seedlings off as soon as they can be handled into a bed of rich soil in a frame, and when sufficiently forward, harden off and plant out in June in well-manured beds in rows of twelve inches apart and eighteen inches asunder. The soil need not be deep, but it must be light, and all lateral shoots must be trimmed off. By the middle or end of September the roots will have obtained their full size, and should then be covered with a little earth to blanch them for two or three weeks previous to lifting them. A portion then may be taken up and stored in sand, but being of a very hardy nature, the bulk may be left in the ground, and if covered with six or eight inches of light soil will be quite safe.

The beds must be kept free from weeds. They should have an occasional soaking with liquid manure in dry weather. They only require earthing up a few inches with the hoe. Celeriac is grown in Dresden to great perfection.

The best kind is Vilmorin's New Apple-shaped Celeriac. A very hard kind, called Knott Celery, is grown in Germany.

## RECIPE FOR COOKING

They should be carefully pared and boiled till tender, then served either whole with melted butter or white sauce, or cut into slices with white sauce poured over them. As a variety, the white sauce can be flavoured with Parmesan cheese.

# CELERY

This favourite vegetable was formerly in a wild state and grew in ditches, and was called 'smallage,' but with cultivation its original state is almost unrecognisable. The first sowing should be made about the end of February, a second sowing the first week in March, and a third sowing about the middle of April.

For the first crop in February, the seed should be sown in a box or pan and kept in a warm greenhouse, and then the young plants transported to a hotbed in a frame as soon as they are fit to prick out. The soil should be rich, they should be frequently watered, and the earth stirred, and in favourable weather plenty of air given them. The second sowing in March should be the same as the first, and the pricking out may be placed in open situations in warm showery weather. The young plants should be put in six inches apart in the nursery beds, well watered, and the earth around them constantly stirred.

The third sowing in April may take place in open frames and attended to as before mentioned.

*Planting out.*—Trenches must be dug about eight inches deep, and the soil should have been generously manured, and should be prepared some time before they are wanted, as they will then be in a better condition than when only dug as wanted. The manure must be rich but not fresh or rank, and it should be well blended with the soil in the bottom of the trench.

Plant in single rows about ten inches apart, and never allow the plants to get dry ; and when the growth is well advanced, give liquid manure twice a week, not in driblets, but a thorough soaking.

*Earthing up.*—When the plants are about eighteen inches high (or less, according to variety) they are ready for the first earthing up. This should be done *most* carefully, as great damage will be done by filling the hearts of each with soil, and perhaps, if too moist, cause decay. To prevent this, tie together the leaves with bast, which is perfectly effectual. First, all the decayed leaves and suckers should be removed gently, as they weaken the growth of the primary root. The earth should be well broken up finely before earthing up, then the earth laid about the stalks of the leaves, firmly pressing it down till the leaf part of the stem is reached. The bast with which the leaves are tied up is slightly buried under the soil ; it should be cut and left. When the celery has made some growth again, it should have another earthing up in the same manner, and as long as the growth continues it should be occasionally earthed up.

Johnson, in his ' Gardener's Dictionary,' gives very minute directions for the earthing-up process. He says: ' The first earthing up should be done with a small trowel, holding the leaves together in one hand and stirring and drawing up a little earth to the plant with the other. The next earthing up is done by the help of two light boards six to eight inches broad, of the same length as the trench is wide, these to be placed between two of the rows of plants ; then place between these boards well-broken earth as required, then draw up the boards steadily, do the same in the next space, and so on till all is completed.'

He says also : ' By this method more than double the quantity can be grown in a given space of ground, and it is handy for protection in winter, either with hoops and mats or litter.'

During the winter celery keeps in good condition if well covered up with long litter, removing it in all favourable weathers.

If there is an appearance of very severe weather three or four dozen of the celery may be dug up and laid in dry earth, sand, or sifted coal ashes, where it will keep good some time. Celery is liable to have its stalks split and canker if the soil is very wet and heavy ; therefore for earthing it up the soil should be light and dry.

The Celery Fly is a great enemy to this plant, and causes part of the leaves to blister and turn yellow, which checks their growth terribly. Beneath the blistered parts small green grubs will be found on examination, which are the larvæ of the Celery Fly, and are mostly to be found from June till November.

The withered leaves should be picked off and crushed so as to destroy the grubs within them.

The ' Cottage Gardener ' describes the Celery Fly as one of the most beautiful of the two-winged flies, and is fond of hovering over flowers and laurel bushes, and resting on palings in the sunshine, from the middle of May to the end of July.

Some of the best kinds are : White Celery, Sutton's Gem ; Sandringham White, and Wright's Giant White for the Red Celery ; Sulham Pink, Manchester Red, and Carter's Incomparable Crimson.

## RECIPES FOR COOKING

### Celery à l'Italienne

Take a couple of heads of celery, well washed and blanched ; stew them in milk till quite tender. Make a brown sauce, and, after draining the celery from the milk, cover it with it ; then put on a layer of grated Parmesan cheese ; then more celery and sauce ; then more; on the top scatter a few breadcrumbs, and lay on little bits of butter over all, and brown with the salamander.

### Celery Soufflé

Take some sticks of nicely bleached celery, chop it, and pound it in a mortar, and then put it into a saucepan with cold water sufficient to cover it. Add pepper and salt to taste, and bring to the boil. Then strain the celery, and put it into a stewpan with half a pint of milk, and cook till soft enough to strain through a hair sieve. Then melt one and a half ounces of butter in a saucepan, and pour it on to an ounce of cornflour which has been made into a smooth paste with a little milk ; add cayenne pepper and salt to taste, and a teaspoonful of lemon juice. Beat the yolks of three eggs till creamy and smooth, and add to the celery ; and lastly add the whites, beaten to a stiff froth. Pour the mixture into a soufflé dish, filling it to the half only. Sprinkle a little coralline pepper and finely chopped parsley over all, and cook in a moderate oven for fifteen minutes. ('The Queen.')

## Fried Celery

Cut the white celery into pieces of about four inches long, dust them with salt and pepper, dip them into well-beaten eggs, then into fine breadcrumbs, and fry quickly in smoking hot fat ; drain, and serve very hot. This is a good accompaniment to roast turkey, pheasant, or chicken.

# CHERVIL

This is much used in salads and mayonnaise sauces, and makes a charming garnish to many of the fashionable cold entrées and savouries. It is also very good to be put amongst the flavourings for soup. The soil should be well drained, light, and chalky. It can be sown in the open ground any time in succession from February till September, and the seeds sown in drills eight inches apart and not more than a quarter of an inch deep ; the seedlings should be thinned out six inches apart in rows. The seeds should be sown on a warm day. A moist and cool situation suits it best.

# CHILI

This is a very tender annual, and must be raised in heat, and instructions for the cultivation of capsicums answer for this. There are three varieties of it—the scarlet, yellow, and purplish black. Sutton's Coral Red and Tom Thumb are the best of the red kinds, Black Prince of the purple.

They are used in curries and many little fancy savouries.

*RECIPE*

## Chili Vinegar, to Make

Take 100 fresh red chilies and cut them into halves ;
let them infuse for a fortnight in a quart of the best pickling
vinegar ; then strain and put it into small bottles ; keep the
bottles well corked. (Cassell.)

# CHIVES

The plant is a native of Great Britain, is peren-
nial, and grows well in any ordinary soil. A
light, rich one suits them best. They are propagated by dividing the roots in spring or autumn.
It is best to buy the roots in the first instance, and
then, in March or April, plant together eight or ten
of the offsets of the bulbs in rows ten inches apart,
and as many from patch to patch. The clumps
should be cut regularly for the maintenance of the
continuous growth of the young roots. Every four
years they should be lifted, divided, and the roots
replanted. They are very much more delicate than
onions, and are much used for flavouring and in
cutting up for salads and omelets.

# CUCUMBERS

The principal points in growing cucumbers are
plenty of heat, moisture, and rich soil, and without
these failure is the result. The temperature should
range from 60° as a minimum to 80,° 90,° or even
100° occasionally, under the influence of bright
sunshine.

A constantly moisture-laden atmosphere is of
quite as much importance as a liberal supply of it

at the root. A dry surface should never be seen. In a cucumber house, paths, stages, and walls ought to be always kept in a moist condition by syringing or by the use of the rosed-can. In hot, bright weather the paths of the house should be deluged two or three times a day. At the same time the bed must be well drained, so that all superfluous water may run away at once, and the plants ought to be set high. Ventilation should be very cautiously given, in order to check evaporation. Moisture rather than much fresh air is chiefly of consequence. As regards soil, they will grow almost in any moderately light, rich, sweet, and porous compost, though nice turfy loam is preferable, chopped up roughly, and mixed with a small proportion of old rotted manure, and, unless naturally sandy, some sharp grit of some kind. A little soot may be added with advantage, and extra nourishment in the form of liquid manure may be given subsequently. 'Gumming' in cucumbers is due to cold, and often occurs in the early part of the season if the spring is dull and cold. If the plant is badly affected it cannot always be cured, but it can generally be got over by raising the temperature and keeping the atmosphere moist.

**To grow Cucumbers in Frames.**—April is a good time to make up a cucumber bed. A bed of warm manure should be made up—stable manure by itself heats too violently—throw it into a heap to ferment, and when it becomes hot it must be turned over and thoroughly mixed and pulled to pieces by placing the outside of the heap in the middle, and *vice versa* ; and in a few days, when the heap is getting hot, turn it over again and water the dry spots if there be any then. Then build up the bed and put the frame over it, the bed being a foot larger all round than the frame ; tread it down well, then put a thermometer into the bed, and as soon as the heat becomes steady between 85° and 90,° put in a barrowful of soil in the form of mole-hills, and arrange one under each light.

The best soil for cucumbers is turfy loam two-thirds, and one-third old manure in a thoroughly mellow condition. Plant a few seeds singly in small pots, and plunge them into the bed ; they will be up in a week, and soon large enough to plant in the lights, one plant in the centre of each mole-hill. Fill up the bed three weeks after planting with loam, to the depth of six inches, and pinch the leader out of the plants when the second leaf can be seen, and peg the young bines, which break away over the surface in such wise that the frame will fill quickly.

The plants will not require much water at first, but when bearing freely they must have liquid manure once or twice a week, according to the weather.

If the soil is moist, as it should be, at planting little water will be needed, except to sprinkle the leaves in the afternoon of each day about 3 P.M. ; the water should be tepid—never colder than the heat inside the frame or house. When the plants are eighteen inches high, pinch out the point of each to induce side shoots to form, and where these are one foot longer, pinch them also, following up this treatment as long as the plants are growing. Remove all weak shoots, and cut away occasionally a few leaves to give light and air to the plants. Air should be admitted freely, yet carefully, to the plants, and foliage should not be cut unnecessarily. If possible, keep up a temperature of 60°, allowing it to run up to 75° with air on. If the nights are cold a mat should be thrown over the glass. The foliage should be sprinkled every day with tepid water when closing the frame ; as the season advances the foliage should be syringed twice a day, early in the morning and again about three. Abundance of moisture is needed atmospherically in the house, as it is not only beneficial to the cucumbers, but for the prevention of insect pests ; a dry atmosphere encourages the spread of red spider.

Ventilation is also most necessary in cucumber cultivation, and should be given whenever conditions are favourable. It is a wrong practice to allow the temperature to rise considerably before putting on air, as is often done, as then whatever ventilation is put on will not counteract the evil done, and consequently the cucumber very quickly gets a parched appearance.

One often sees cucumbers form freely, but instead of coming to maturity they damp off, and direct light is needed to form a fruitful growth.

Cucumbers require a great deal of *feeding* ; if not, innumerable insects will attack them, as well as mildew. To feed them, give them manure water made from fresh cow manure and soot, which must be applied in a diluted and clean state.

If the cucumbers are in a house or greenhouse, it is a good plan on closing it on the evening of fine days to damp the floor. The house or frame should be shaded on bright days from 10 A.M. till 4 P.M., then sprinkled and closed. The frame should be opened just a little in the morning by seven o'clock, increasing the ventilation according to the temperature outside.

A layer of moss litter manured is a great source

of nourishment to the plants all through the season.

Sutton, in speaking of cucumber culture, says : ' In warm sunny days two or three syringes of water will be beneficial, but must not be done so late as to risk the foliage being wet when night comes on. On dull days one good sprinkling over the foliage will suffice, which should be done in the morning ; and there will be occasions when it may be advisable to avoid touching the leaves with water, if there is no probability of their drying before nightfall ; and in the event of the bed falling below the proper temperature, the water may be a few degrees higher than usual. It is also necessary that the interior of the glass should be frequently wiped to prevent the condensed steam dropping on the plants, which is very injurious to them. Where the plants have not the benefit of currents of air or bees to convey the pollen of the barren plants to the stigma of the fertile ones, the latter must be dusted by the grower, or the plants must be exposed to as much air as possible in the middle of the day when it is warm enough during the time that they are in flower.'

Cucumbers should be grown as long and as straight as possible.

**Ridge, or Out-of-door Cucumbers.**—Sowings must be made at the close of May or early in June. It is better to raise the seeds in a cool greenhouse or frame, unless there are hand-lights to put over the seed as soon as sown. A warm border is needed, and they can be grown on level ground if it has been well manured and dug up a good spit deep a few weeks before planting time. The plants require a little protection when first set out by turning a large pot over them for a few hours during the day when the sun is very bright, also at night, until the plants get too large for the pot to cover them ; and grown in this way the plants do not require watering unless the season is hot and dry, but still they must not be allowed to suffer from draught, and if it is necessary to water, the water must be soft and well warmed by exposure to the sun, and then water liberally three or four successive evenings, and then give no more for a fortnight.

Sutton says ' the plants should be only pinched once, and there is no occasion to set their bloom, and let them grow as they please.'   If good, strong plants are set out about the middle of May there will be plenty of cucumbers to cut in July if rightly managed.

Where cucumbers which grow in frames are desired straight, it is a good plan to utilise cracked lamp chimneys by placing them so that they grow into them, though to purchase a few new ones would not be ruinous, and would greatly improve the appearance of the cucumber.

**Premature Decay of Cucumbers.**—There are various causes for this, such as over-cropping, dryness of the roots, or stagnation caused by too much moisture at the roots, or insufficient drainage ; also fluctuation of temperature will cause the young fruit to turn yellow.   Another cause may be, if in a cucumber house and no fire heat is employed on a cool night succeeding a hot day, the plants receive a check which causes a stoppage of growth and the fruit to decay.

A very troublesome disease which occasionally attacks cucumbers is that warty excrescences form on the roots, which cause the plants to become sickly and the fruits die off. For this there is no cure, and the only thing to do is to pull all the plants down and burn them, take away all the old soil, and clean the house out and start afresh with fresh soil and new healthy plants. When cucumbers are found to be bitter, it is because they have not had enough water to the roots. Gumming is usually induced by low temperature and stagnant roots, for which more heat and stimulant should be given.

Cucumbers are very liable to be attacked by Green Fly (*Aphis*), Red Fly (*Acarus*), and Thrips, and for their expulsion see the supplement, 'Garden Pests.' They are also subject to mildew, canker, and gumming, for the prevention of which see supplement, 'Diseases of Plants.'

**A few Hints for Cucumber Growing.**—A constant supply of moisture is most necessary. In addition to liberal applications of water or liquid manure at the root, the whole of the interior of the frame or house for cucumbers must be thoroughly syringed twice a day in bright weather, and even three or four times on very warm sunny days, especially if there is much wind. The foliage should be shaded to prevent its flagging, but no more. Plenty of moisture will keep it up under all but the strongest sun. Close the frame an hour or two at least before the sun goes off it in the afternoon (after a good damping down), and let the thermometer rise to 70° or 100° for a time. A little air may be given later on if necessary.

The best kinds of cucumbers to grow are Telegraph, Lockie's Perfection, Sutton's Duke of Connaught, and Hutton's Magnum Bonum. For Ridge Cucumbers: Sutton's King of the Ridge,

Stockwood, Athenian Ridge, and Short Prickly (or Gherkin) for pickling.

**Winter Cucumbers.**—The seeds should be sown singly in 3-inch pots placed near the glass. The best varieties for the winter are Telegraph and Cardiff Castle, the preference for the first.

The soil in which the plants are to be sown should consist of one half loam and the rest of decayed manure (that from an old hotbed preferable) and leaf soil. The compost must be well mixed and placed in the house four or five days before planting, so that it may be warm quite through. The glass and woodwork should be thoroughly cleaned previously, both to give light to the foliage and destroy any insect pests.

When the plants are to be turned out of the pots, care should be taken that the soil is thoroughly moist. They may be syringed twice a day except on very wet days. To check red spider and thrips, as soon as the fruit commences swelling liquid manure water in a tepid state should be applied to the roots : this should be very weak and applied often.

The temperature should range between 65° and 75°, but on bright sunny days with a little air it may be 5° higher ; in fact, with a little extra fire heat the ventilators might be open every day. The atmosphere as well as being warm should also be moist, and for this purpose water must be thrown on the floor occasionally, but not too often, because in the winter the leaves are liable to be attacked by mildew, which, should it appear, must be quickly destroyed by dusting with sulphur.

If too many fruits appear, all but the best

should be removed. The shoots must be stopped at every second point when the trellis is sufficiently covered, and some of the old wood can be cut out to allow the young growths room to develop.

**Cucumbers Failing.**—Canker is a disease to which cucumbers are subject. Too much moisture just round the stem, where there are few or no fibres to take it up, is a frequent cause of canker ; therefore the collars of the plants should be kept high. Not to wet them there more than can be helped, it is a good plan to place some pieces of charcoal or coal cinders round each.

Canker, if not taken in time, generally proves fatal. To grow cucumbers really well they ought to be pushed on in a genial temperature from $70°$ to $80°$, and never falling below $60°$ at the lowest, with plenty of moisture both at the roots and the atmosphere, and nourishment as well. But little ventilation is necessary except in very hot weather.

*RECIPES FOR COOKING*

## Concombre Farcie

### Stuffed Cucumber

Remove the peel from a large cucumber, cut off the stalk end, and scoop out the pips and a portion of the inside. Have ready a forcemeat made of meat, poultry, or fish, with finely chopped onions and chervil, and fill up the inside of the cucumber with it ; envelop with slices of bacon tied over, and put into a stewpan with some nicely flavoured pale stock and a bouquet garni, and let them simmer till tender. Now remove the cucumber, and keep hot whilst the sauce is boiling to reduce it ; thicken it with flour and butter in the usual way, pour it on to a dish, lay on the cucumber, and sprinkle with lemon juice.

## Cucumbers and Eggs

Cut up a cucumber into dice-shaped pieces, and put them into boiling water. Then take them out and put them into a stewpan with an onion, a good-sized piece of butter, and a piece of pork ; add a little salt. Cook with cover on for about fifteen minutes, sprinkle then with flour, add cover with good veal or chicken gravy, stir well, and keep over a gentle fire till no scum rises. Then remove the onion and pork and add the yolks of two eggs and a teaspoonful of cream. Stir, then remove from fire, and squeeze in a little lemon juice. Send to table with poached eggs on the top.

## Fried Cucumbers

Pare the cucumbers, cut them into very thin slices, season with salt and pepper, then dip them into beaten egg, and then into breadcrumbs. Put a couple of tablespoonfuls of good dripping into the frying-pan, and when hot put in a few slices of the cucumbers ; when brown and crisp on one side, turn and brown the other. Take out carefully, drain, and serve hot.

## ENDIVE

Endive originally came from China and Japan, and was known in England before 1548. It is most easily produced on freely manured light soils, and nothing is so good for it as growth in the soil cleared of early potatoes.

Where there is not too much room in gardens the seed should not be sown till the middle of April or May. The seed should be sown in drills twelve inches apart, and not very deeply in the surface, and when the plants are an inch high prick them out on a bed of rich light soil. The plants in the seed-bed should be thinned to three inches, and must be watered in dry weather. When planted they should be set in rows twelve inches apart. Water must be applied every evening till the plants are established.

Much endive is lost because so many growers blanch too much at one time, or omit to store the bulk of the crop out of the reach of frost. The plants decay very quickly after the tips of the leaves have been crippled by frost.

The safest plan is to grow the endives in beds, covered by lights or mats whenever frosts are expected. Often endive is spoilt by either blanching too much at a time, or because the bulk of the crop has not been stored out of reach of frosts. Endive when lifted carefully will keep wonderfully well in sheds, and when there is but little space it can be stored in conical heaps surrounded by dry sand, the points of the leaves being well brought together so as to protect the hearts facing outwards.

*To Blanch Endive.*—Endive keeps badly after it is blanched, therefore very little should be done at a time. This operation should not be begun till the plants are fully grown, and as long as the plants can be left in the open, the blanching can be most simply effected by tying the outer leaves well together, excluding the light from the hearts ; they can also be enclosed singly in 6-inch or larger pots inverted over them and the holes stopped ; also slates or tiles may be laid on them.

Endives should be kept well moistened at the roots, as dryness causes the leaves to become tough and also induces premature flowering. It keeps wonderfully well in sheds, or it may be stored in conical heaps surrounded by dry sand, the points of the leaves being well brought up together so as to protect the hearts.

The best kinds to grow : Sutton's Incomparable Green, White Curled, and Moss Fine Curled. For winter use : Round-leaved Batavian, which is very acceptable when lettuces are scarce.

## *RECIPE FOR COOKING*

### Endive au Jus

Wash and clean the endive, cut into halves, season with pepper, salt, and a grate of nutmeg. Fasten the halves together again, and place them into a saucepan, with a slice of bacon over and underneath it ; add two onions stuck with two cloves, two carrots, and a bouquet garni ; cover with some good gravy, and let it simmer till tender ; then give a squeeze of lemon over, and send to table with poached eggs on the top.

## FENNEL

This herb should be sown in March to May in light warm soil in drills, and transplanted when three or four inches high. A sandy soil suits it best, but not wet ; the plants should be planted when thinned to eight or nine inches apart. Fennel is used generally for making fennel sauce, and eaten with mackerel ; it is also used in soups, and for garnishing it is most effective. The seeds are also used in medicine.

# GARLIC

This should be grown on a light rich soil, and should have a place if possible in every garden, for its use in flavouring is so necessary, as a soupçon of garlic in rumpsteak pudding, hashes, &c., is such an improvement. It is best sown in a light rich soil, and it is generally propagated by parting the root, and it may also be raised from the bulbs produced on the stems. The middle of March is the best time to plant. They should be planted about two inches beneath the surface of the soil, and about eight or nine inches apart ; holes should be made in straight lines, and a single clove or bulb must be put into each, the roots downwards. They must be frequently hoed, and the leaves tied in knots in June to prevent the plants running to seed. They may be lifted as required in June and July, but the *whole* must not be lifted till August. In lifting them, leave a piece of the stalk attached so as to tie them in bundles with it after drying them for storing for the winter.

Garlic is a very good thing for driving grubs, snails, and moles away. The bulbs should be slightly crushed and then strewn about.

# KALE (SCOTCH)

This vegetable is of a very hardy nature, and in cold and exposed places it is indispensable. There are a great many varieties of this plant, but are pretty similar as regards their cultivation.

They delight in rich, well-worked soils with copious manure waterings.

The seeds should be sown broadcast and thinly in about three successive sowings, commencing the end of March, then in April and May ; the seedlings should be pricked out when their leaves are about two inches in width, and set about six inches apart each way, and watered frequently. In a month's or five weeks' time they will be ready for their final transplanting. They must be kept well weeded, and the earth drawn about the stems of the large spreading growth.

The most useful kinds are the Dwarf Curled or Scotch, the Cottagers, the Jerusalem or Asparagus Kale, and the Phœnix ; Variegated Borecole or Garnishing Kale ; Chou de Milan, which is a hybrid savoy, and Chou de Burghley, a cross between a cabbage and a broccoli, useful in large gardens, and Purple Borecole ; but Curled Kale, Buckman's Hardy, and Variegated Borecole for preference, the latter being most useful for garnishing.

### RECIPE FOR COOKING
### Kale à la Montglas

Cook the kale in the usual way. Chop it, put a little butter, pepper, and salt to it, and form into the shape of large leaves, and let one overlap the other, placed round a dish like cutlets.

## KOHL RABI

This is a most acceptable vegetable, and should be grown in every garden, however small. It is grown as a substitute for turnip, being much hardier and more delicate and nutty in flavour, besides being sweeter and more nutritious.

This plant partakes of the character of the cabbage and turnip. The soil should be heavy and the seeds should be sown in April and May, then thinned out to three inches apart as soon as possible. Then they must be thinned again, and planted two feet apart in rows ; they must be planted shallow, as the plants must have the chief part of their stems left uncovered by the soil. The hoe must be used every now and again to keep the ground clean and ventilated. In appearance it is globular in form, very like a large Swedish turnip, with stems crowned with leaves scalloped on the edges.

It is a most useful vegetable, as the green part can be either cooked as cabbage or spinach, and the roots boiled like turnips.

There are several varieties, but the green-stemmed and the purple-stemmed are preferred. The best kinds to plant are Large Green, Large Purple, and Neapolitan Curled Kale.

### RECIPES FOR COOKING

Simmer a kohl rabi till tender, and then pare and cut it into slices after draining it. Then place some slices into a china fireproof dish with a little good white sauce, into which some grated Parmesan cheese has been mixed, and serve very hot.

## Kohl Rabi Curried

This is made as in the former recipe, only instead of the white sauce a curry sauce is used.

## LEEKS

The leek originally was a native of Switzerland, and was introduced into England before the time of Elizabeth. It belongs to the onion tribe, only far more delicately flavoured than onions. They require a long season of growth in a deep, well-worked, highly manured soil.

Sow about the middle of March in the open, and, when large enough, they must be thinned and transplanted as opportunities occur—which, as Sutton recommends (and I have found it the best plan), should be when the plants are six inches high—into well-prepared and previously watered ground, and trenched in a similar manner for celery, eighteen inches wide. This will allow two rows of plants being put into each trench. Manure should be laid at the bottom of the trenches. Cow-manure is best for light soil, and that from the stable for heavy land.

The plants should be put in one foot apart in the rows and fifteen inches between the rows. Stir the ground occasionally, and cut off the tops of the leaves to cause the roots to grow to a larger size.

Once a week in dry weather they should have a good soaking, and early in September, when a good growth has been obtained, the leeks should be carefully earthed up so as to have them thoroughly blanched. It takes about six weeks for them to blanch thoroughly. They should be covered with the earth nearly down to the leaves.

The tops of the leaves should be cut off about once a month, and, as new ones are produced, the neck swells to a much larger size. It is a good plan, as soon as winter sets in, to take the best roots and bury in dry sand, and they will keep good for from six weeks to two months at least.

Sutton advises that as the flower stems rise up *every* one should be nipped out, which will result in the formation at the roots of small roundish white bulbs, which make a nice dish when stewed in gravy, and may be used in cookery the same as onions, and which are to be obtained only in the spring or early summer. We have not tried this part of the cultivation of leeks yet, but intend doing so.

The best kinds to grow are Sutton's Improved Musselburgh, Large Rouen, Ayton Castle, London Flag, and Poitou. The Lyon is a *very large* variety.

### *RECIPES FOR COOKING*

## Flammish Leeks

Wash and blanch some nice-sized leeks, cut them into half-inch pieces, put half a pint of cream into a basin, and stir in the leeks ; season with pepper and salt. Have some short paste, and divide into four parts ; roll out each into rounds about the size of cheese-plates ; place the leek mixture into the centre, and gather up the sides of the paste like a puckered purse ; fasten with a small round piece of paste, wetted, and fixed on the centre. (Cassell.)

## Leeks au Gratin

Prepare the leeks as above, then place them in an oven, and as they get hot put bits of butter on them, and dredge them lightly with biscuit raspings ; when these become a nice brown add some more, till the leeks are tender and well coated with the raspings. Sprinkle with cayenne pepper and serve with lemon cut into quarters to be squeezed over them.

## LETTUCE

The lettuce was introduced into England in 1562, but its origin is not known, except we hear of it in the East about the time of Cyrus the Great.

The soil should be well dug and richly manured, though lettuce will grow almost in any soil, and may be planted between other vegetables, especially in rows between cabbage, which is of course a great economy of space where the ground is limited. The situation *must* be sunny, they must have an abundant supply of moisture throughout their whole growth, as the more water is given the less apt are the plants to run to seed, besides ensuring crispness and succulency.

Sowings should be made in boxes in February and put into frames, after which time they can be sown in the open. When sown in frames they should be pricked out when three weeks or a month old, thinned to three or four inches apart, and prick out those removed to similar distances; those from the February sowings in frames, and thence until August in any open situation.

In the final planting out place them in rows a foot apart each way. Lettuces always attain a finer growth if left where sown, and those which are planted out at once are in every respect better than those pricked out for a final planting.

Those required for winter use are best planted on ridges, and can be sheltered with hoops and matting during severe weather. Occasionally an application of liquid manure is most beneficial, but watering liberally must take place throughout their culture.

There are two kinds of lettuces, the Cos and the Cabbage.

The Cabbage lettuce grows flat and spreading, and the Cos compact and upright.

Cos lettuces are generally blanched by bending down the tips of the leaves over the heart and tying them together with bast. Johnson recommends in his 'Gardener's Dictionary' that at the time of tying them up, the centre bud of each should be cut out with a sharp knife, to prevent the plants running to seed before the heart is perfectly blanched.

The German or Spotted lettuce is being now much grown in England. It is cultivated in the usual way, and is similar to the Cabbage lettuce, only larger, and with reddish spots over the leaves. It is very succulent and sweet in eating.

The Italian lettuce Incappacciato can also be well grown here, and is a very crisp lettuce cabbage in form, but with jagged prickly-looking leaves ; they should be quickly grown and eaten quite young. As they grow older they can be dressed as spinach and in various other ways, and make a very nice addition and change to the dinner table. We have grown it with seeds sent from Naples, and found it succeed wonderfully well.

The best kinds of English lettuces to grow are : Cos Lettuce, Sutton's Mammoth White, Paris White, Winter White, Cabbage Lettuces, Sutton's Gem, Commodore Nutt, Golden Ball and Veitch's Perfect Gem.

*RECIPES FOR COOKING*

## Dutched Lettuces

Wash a couple of lettuces, separate the leaves, and tear each leaf into two or three pieces ; cut a quarter of a pound of bacon into dice, and fry till brown ; whilst hot add two tablespoonfuls of vinegar. Beat an egg till light, and put to it two tablespoonfuls of sour cream, mix it with the bacon, stir over the fire for one minute till it thickens, and pour boiling hot over the lettuce ; mix well, and serve quickly.

## Lettuce à la Milano

Take some lettuces—the Cabbage, Spotted, or Italian Incappacciato are best—boil as for spinach, run through a tammy, and finish with flour, butter, a little milk and sugar, the same as for spinach.

## MARJORAM

There are four kinds of this herb, the Common, the Pot, the Sweet, and the Winter variety.

The first and last are hardy, and should be increased by dividing the roots in spring.

Sweet Marjoram is grown as an annual by sowing the seed in April on a warm border, and subsequently transplanting the seedlings about six inches apart.

## MINT

The best to grow for all ordinary purposes is the true spear mint. This herb is increased by division of the roots in spring and covering them with two inches of soil. It requires rich and moist ground. Cuttings of the young shoots root freely in the early summer if kept moist and shaded. It is very necessary to keep the mint bed clear from weeds, and to allow plenty of space for growing. To have mint ready for Easter, some roots should be put into frames or some planted in a box and kept in a greenhouse.

Peppermint is grown in the same way.


## MUSHROOMS

The mushroom was one of the table delicacies in the early days of English history. From some of the Roman records we read that the Emperor Claudius met his death at the hands of his wife Agrippina (who was also his niece), who had prepared him a dishful of the poisonous species. In France mushrooms form a very large article of consumption, and beds of them are cultivated frequently miles in extent. A cave at Méry contained in 1867 twenty-one miles of beds, producing not less than 3,000 lb. daily. The catacombs and quarries, Moulin de la Roche, Sous Bicêtre, and Bagneuse produce immense quantities of them. These are all under Government supervision, and are periodically inspected.

Mr. Henry de Vilmorin writes that mushrooms can be produced easily everywhere, and in all seasons, with a little care. The essential conditions are to lease a very rich soil and a uniform temperature, which should be kept from rising above 86° or falling lower than 50° Fahr.

To construct the bed, although horse-dung is generally employed for this purpose, all other warm manures are suitable, such as those of goats, sheep, rabbits, or fowls. The manure should not contain too much straw, neither should it be too close or too highly charged with ammonia.

The dung, whatever it is, should undergo a preparation to moderate the fermentation, which will at the same time make it more durable and equal. As soon as the dung has been taken from the stable it should be put into square heaps about one yard in height, and all foreign substances removed. It should be well mixed with a fork.

The parts which seem to be dry should be moistened, and then trim or press firmly together the sides of the heap. It is to be left in this state until the heat becomes excessive, which condition will be known by the white colour of the most odorous parts, which takes place in from six to ten days after the heap is made. The heap must then be broken down, the manure shaken up, and the heap rebuilt, taking care to place the manure that was outside, and of which the fermentation is the most backward, in the inside.

Generally, some days after the heap has been turned, fermentation will be so great that the heap will have to be thrown down and remade.

Sometimes, however, after the second working

the manure is sufficiently made, and can be put into beds. When the manure has become brown without being rotten, it can be used without danger, for then the straw has lost its consistency, and its smell will resemble the odour of fresh mushrooms.

A heap of *less* than a cubic yard can hardly be handled properly.

The best way of constructing the beds is to give them a height from nineteen inches to twenty-three inches, with the width nearly equal at the base.

When a large space is to be laid out, beds in sloping form or with a shelving edge are preferable, as they can be of unlimited length whilst keeping the height and breadth named. The breadth, on the contrary, ought to be less than the height when the beds lean against a wall. Beds can be made in old tubs sawn in two or upon planks. In the latter case they have given to them the shape of a cone, or more like a heap of stones which one sees by the roadside. The manure ought to be easily divided, and when putting it in its place the parts which form clods should be broken, and the compact portions mixed with the strawy parts and all well stirred together. The manure should be trampled over three or four times, and then all projecting pieces removed, so that the surface is firm and smooth.

After the beds are made it will be necessary to wait a few days before putting in order to see whether fermentation begins again.

Use a thermometer, and if the temperature is above 86° Fahr. the bed is too warm ; then aerate the bed by pushing a stick into it, so that the heat can escape from the holes. As soon as the excess of the heat is over, the heap must be rammed down again. When the temperature is uniformly about 77°, the spawn can be put in. Either fresh or dry spawn can be put in. Some days before introducing the dry spawn into the beds it is good to expose it to the influences of a tepid and moderate moisture, as it makes the growth quicker and surer. To impregnate the beds the cakes of spawn are broken into pieces three inches square and half an inch thick, and should be inserted in the face of the bed lengthwise, and at intervals of nine inches to ten inches.

In beds which are from fifteen inches to eighteen inches high, and which are the most ordinary size, two rows of spawn are placed.

The pieces of spawn should be put in with the right hand, whilst with the left the dung is raised to make places for them.

They ought to be pushed in till their outside edge is level with the top of the bed, and then the dung must be pressed firmly round them to keep them fixed.

If the bed is in a place where the temperature is equal, it is not long before the spawn grows. If the beds are in the open air, it is necessary to cover them with some long straw dung.

The spawn ought to begin to run in seven or eight days after it has been put in, and it is well to make sure of its growth at this period, so as to be able to replace the pieces of spawn which should not have grown. These are easily recognised by the absence of white threads in the dung. In fifteen to twenty-one days the spawn ought to have spread all over the bed, and should begin to rise to the surface ; the beds should then be smoothed down, beating them with care and to clear up anything which would affect the growth of the mushrooms. The tops and sides of the beds must be covered with earth, which should be light and fresh, rich in saltpetre, and contain some proportion of lime. Old plaster broken up is good for this purpose, and should at first be watered with a weak solution of saltpetre or liquid farmyard manure.

Before putting on the soil, the heap should be lightly moistened, and the earth put on the top and sides to the thickness of half an inch, and pressed firmly against the dung, and made very smooth. After this, all that is necessary to do is to water from time to time, or to replace the straw covering if it becomes removed from the bed. The surface of the bed should remain fresh and moist without being too damp.

If the waterings, which ought always to be light, are not sufficient, the soil round the base of the bed should be watered, which will absorb the necessary moisture.

Some weeks later mushrooms will begin to appear. When picking the mushrooms the holes left should be filled with the same earth which has been used to cover the bed. The production extends generally from two to four months if left to itself, but it can be maintained longer by means of lightly watering, the water having guano added to it. The temperature of the water should be from 68° to 86° Fahr., but great care is necessary in watering, so as not to injure or dirty the growing mushrooms, as through neglect of these precautions mildew and rottenness will often occur.

Pull up all mushrooms which appear soft and yellowish, throw out the earth which is around them, and water the holes from which they are pulled with a mixture of three-quarters of an ounce of saltpetre in a pint of water, and refilled with earth. If small white granules appear at the bottom of the bed and round the pieces of spawn, take out the affected parts and refill with fresh earth.

Besides raising three or four lots of mushrooms under cover in the year, the supply can be continued out of doors during the summer season. Beds used for other forced culture may be pierced with a stick in their sides and mushroom spawn placed in them, and they will often yield good crops, provided that the temperature is proper and the young mushrooms are protected by a light covering of earth when they are beginning to develop.

To test it, plunge in an ordinary thermometer. The spawn should be broken into pieces two inches square. Insert them into the manure two inches deep and nine inches apart all over the bed. Press the manure down firmly, and just cover the bed with fine sifted earth about an inch thick, and press down firmly with the back of a spade. The heat of the bed should never be allowed to get lower than 65° for the first six weeks. It is a good plan to do this by adding a little straw on the top.

**To make Mushroom Beds in the Open Air.—** Proceed to make the bed as above. Beds for winter bearing should contain more material than for the summer work. Straw manure is best, and should be procured from stables where the horses are receiving *hard* food, and only the longest of the litter should be shaken out, and which can be used for covering the beds when dried in the sun.

The beds for winter should not be less than three feet wide and equally high, or nearly so. Soil up to two inches in depth should be placed on the beds when the spawn is running freely. Choose a sheltered spot for an outdoor bed. Mushrooms for autumn and winter should have a bed the south side of a wall or hedge ; if for spring and summer, behind a north wall. Should the heat of the beds become too hot, make a few holes down the centre of the bed with an iron stick or bar.

Mushrooms ought to appear in six weeks, but that cannot be depended on, and sometimes are much longer in producing them. Outdoor beds if well covered seldom require watering, but if they become *dry* they must be moistened with lukewarm water.

The usual time to water mushroom beds is immediately after gathering. People often make the mistake of not watering enough when the mushrooms are getting forward, as mushrooms always grow best in the fields after rain ; and, when artificially grown, warm liquid manure may be given frequently to beds in bearing.

Mushrooms may also be grown on lawns and in pleasure grounds by procuring some mushroom spawn in April or May, and by breaking the bricks into pieces about two inches square and burying them at intervals all over the lawn. In each place raise a small piece of the turf and insert a piece of the spawn, and press the turf down over it ; then the lawn should be rolled, and the mushrooms will appear in September. The same spawn will produce plants for several years in succession, and the lawn may be mowed just as usual.

The following is of the best plans to make a mushroom bed : First take some fresh horse-droppings from a stable where the animals have been highly fed, and place under a dry shed ; then turn it about occasionally, so that all scent is lost; next place it in a room or cellar, and make a bed ten inches thick; ram the manure firmly, and cover it with one and a half inch thickness of soil, made firm and even. When the bed has attained a falling temperature of about 65°, insert the spawn, which should be put in rows about a foot apart and eight inches in the rows. Each piece of spawn may be put in one inch deep; a little water should be occasionally given. The time it takes for the mushrooms to come up depends upon circumstances—possibly in about six weeks. A good temperature one month with another is from 50° to 54°.

Another way of making mushroom beds is to make them on a brick floor, which should be made in ridge form, as a greater surface is provided than in any other form. The bed should be two feet six inches wide at the base, two feet ten inches high, and nine inches wide at the ridge. Collect some stable manure and throw it into a heap, and protect it from the wet. When enough has been collected for a bed (it matters little how long the bed is made beyond six feet), throw it into a close heap to ferment; if too dry, just wet it very slightly to induce heat. The manure should be turned for three days successively, when it ought to be just at the right moisture to make up into a bed. The bed should be rammed down as hard as possible to make it firm. Fermentation will now set in after a few days, and when the heat is at 85° it will be ready for spawning.

I feel sure mushroom culture would prove lucrative to ladies who are endeavouring to supplement small incomes, if they could manage space enough from their gardens to cultivate them in, as mushroom cultivation is not a difficult art. I read

in the 'Caterer' that a Mr. Joseph Nepp, of Leipsic
and Plagwitz, Germany, who is a civil engineer, has
patented a system for mushroom cultivation.  His
beds are artificially prepared in such a way that
indirect pulsation and aspiration are provided and
the growth stimulated by the chemical bed and the
regulation of the temperature.  These beds can be
placed in cellars, sheds, special houses, or any convenient locality.  They are placed at various angles,
so that a ground space of ten yards square would
give a bed space of fifteen square yards, and ground
space of twenty square yards a bed of thirty to thirty-
five square yards, and so on.  These beds require
practically no attention, and they will go on bearing
a plentiful daily crop from six to ten months, and
instances of beds bearing for twelve to fourteen
months are not uncommon.  The first outlay of a
permanent bed of twelve yards square with a plentiful supply of plant stimulants costs 2*l*., whilst beds
containing 250 square yards cost from 10*l*. to 15*l*.

As mushroom beds decline they must be reno-
vated by taking off the earth, and if the dung is
decayed they must be remade, and any good spawn
which shows itself preserved ; but if the beds are
dry, solid, and full of good spawn, a fresh layer of
compost three or four inches thick must be added,
mixed a little with the old, and beaten solid again.

It is much better to purchase the spawn than
to collect it.  Sutton supplies most excellent and
reliable spawn, and I should recommend everyone
to purchase spawn, as that which is collected from
stable dunghills is not to be relied on.  Sutton has
spawn especially manufactured or them, and it is
so cheap that it seems waste of time to try and
collect it.  There are three grades of mushrooms—

buttons, cups, and broilers—-each used for different culinary purposes. 'Buttons' when the cap is united to the stalk and the gills are not visible; they vary from the size of a nut to that of a walnut. 'Cups' are buttons in a further advanced stage, just showing a ring of the gills half an inch in diameter; and 'broilers' when they are fully expanded, showing all the gills plainly. A little book, called 'Mushrooms for the Million,' published at the 'Journal of Horticulture,' by J. Wright, is a little book that should be in the hands of every grower of mushrooms who means to make their culture pay.

One of the largest mushrooms that has ever been seen, weighing not less than twenty-nine pounds, was lately on view in the window of a small restaurant in the Rue St. Antoine, Paris. It grew to this size in forty-eight hours in the quarries of Porcheville, which are devoted to the cultivation of mushrooms in the Seine-et-Oise Department.

Slugs and snails are great enemies to mushrooms, and often at night may be discovered with the aid of a lantern, and then caught. J. Wright recommends brewers' grains or bran being placed in heaps near the beds, as, if examined after dark, they will be found covered with snails, and if these be covered with salt they will do no further damage.

Wood-lice are also troublesome. If pieces of parsnip boiled in a solution of arsenic are placed in flower pots close to the beds, the wood-lice will eat them greedily, but of course these must not be put where there are fowls, &c.

When mushrooms are the size of buttons, and cease swelling and rot, a fungus has taken possession of them, and can only be destroyed by clearing the beds and cleansing, limewashing, and disinfecting the house, but it is not prevalent in the open air.

Millipedes are small threadlike creatures that sometimes infest the beds, showing that the manure has not been sufficiently heated and purified. Sometimes very minute insects infest mushroom beds, especially if the manure is too dry, and generally one or two ounces of salt dissolved in a gallon of tepid water will destroy the pests.

Mr. Charles Bateson says : ' If sugar and plaster of Paris are mixed together and strewn about an inch thick all round and on the top of the bed, the insects will speedily disappear.'

**To tell if Mushrooms are of the edible kind.**— Wash them in vinegar and water, and wipe them dry and put them into a saucepan full of cold salted water, to which a peeled onion has been added. If during the process of cooking the onion remains white, the mushrooms are fit to eat ; but if it becomes black or slightly discoloured, they are poisonous.

The best plan to remove all peril from using them is to examine every one that is *gathered*, and if doubtful throw it away. Real mushrooms, says Worthington Smith in his little treatise, ' Mushrooms and Toadstools,' are known by their beautiful pink gills not reaching the stem, which stem carries a well-marked white, woolly ring, by the very fleshy down-covered top, the delicious and enticing fragrance, the firm white flesh sometimes inclined to pink when broken.

## RAGOUT OF MUSHROOMS

### Hachis aux Champignons

Take two dozen mushrooms, dry them, and put them into a pan with a piece of butter. As soon as the butter has melted stir in a tablespoonful of flour, two glasses of beef gravy, salt, pepper, and a bay leaf. Cook till nearly reduced by one-half, and then pour it over a dish of hashed mutton. Mix well, and serve with small crusts of fried bread. (Dr. Cooke.)

### Mushroom Creams

Stew some mushrooms in butter, flavour with pepper, salt, and lemon; rub them through a fine wire sieve, and mix with the purée two tablespoonfuls of liquefied aspic jelly and half a pint of whipped cream; thoroughly mix, and fill some small dariole moulds with the mixture and lay on ice. When firm turn out, and garnish with chervil and small cherry tomatoes.

## MUSTARD AND CRESS

It is almost useless to write anything on this small salad, as its culture is so easy, but a book on vegetable culture would not be complete without it. They will grow in any soil or situation. Mustard is the native white mustard in its seed leaves, and cress is a cruciferous plant introduced before 1548, but from what country is not known. The sowing may be made throughout the year. From November to March it should be sown in boxes and put into a greenhouse, and from April to the middle of September it may be sown in the open ground. A change of soil must be made occasionally. The seed should be sewn thickly, and the earth which covers it should be fine and thinly scattered on, for that planted in boxes it is best not to cover at all. Water must be given in dry weather. Sowings should be once a fortnight, and as cress is much slower in vegetating than mustard, the former should be sown five or six days earlier than the latter, so that they will be ready together. Eat it as soon as it is ready, when it is tender, green, and short, and before the rough leaves appear.

## ONIONS

The soil which best suits onions is a sound, deep, and rich soil, yet moderately free, light, or sandy loam. The soil for planting them should be in a fertile and well pulverised condition ; a rather gritty soil makes the best seed bed, and where there is much clay other matter should be introduced into the soil. Coarse sand, burned refuse, are the best materials to favour successful growth. They also require liberal manuring. The ground should be well trenched and manured in the autumn, throwing up the surface in rough ridges, so that frost may act upon it thoroughly during the winter, and then in the spring it will break down beautifully, and then should be levelled and raked to a fine even surface and the seeds sown, which should be done during March, again in July and early in August for autumn growing, also early in September.

Sow thinly in drills eight inches apart, and the beds should be about four feet wide, and after the seed is raked in, the surface of the bed should be beaten flat with a spade. In about three weeks the beds should be hoed and thinned out, and these should be then young onions fit for salads. The beds should be again hoed and thinned out from time to time as the onions may be wanted. Carefully remove all weeds, as they are very injurious to the crops, and let the air circulate freely among them. Onions sown in March will be ready to take up in August or September. It is a good plan about a month or six weeks before the onions are ready to take up, bend the stalk down flat on the bed, to throw all the strength of the plant into the bulb, and to prevent its thickening at the neck ; the bend should be made two inches from the neck. Watering is not often needed except after transplanting, which must be given for several successive evenings.

A little liquid manure given twice a week is most beneficial to their growth. Watering the bulbs after they have attained to a reasonable size will do more harm than good.

When onions are ready to take up, they should be drawn and dried on the ground, about ten days after which they may be gathered together and topped and tailed, and made into bunches or ropes.

Onions of the largest size are obtained by sowing the seeds in August and pricking out the seedlings from six inches to one foot in the early spring. In storing onions, any dry cool place will answer ; they may be hung in bunches on an open wall, under the shelter of the eaves of any building, but during frost they must be taken down and put into a *dry* place under cover ; damp starts them into growth, when they will have topped and tailed again. Onions may also be placed in heaps in dry outhouses and well covered with matting, or simply laid in a thick layer on shelves, or on the floor of a dry cool room or shed.   Properly stored onions do not decay readily, and if dry are not injured if exposed to a few degrees of frost. Onions are particularly liable to ravages from the Onion Fly (*Anthomyia cepa*) and the Brassy Onion Fly (*Lumerus*).   These pests cause the young onions when quite young to turn yellow and the leaves sink down on the ground, and the white portion is pierced to the centre by a fleshy shining grub a quarter of an inch long.

The best remedy is to sprinkle gas-lime between the seedlings, as its fumes are offensive to the fly. A regular use of salt and soot is an effectual preventive, and also spreading powdered charcoal among them, as the fly will deposit her eggs on it instead of the plant.   The larvæ will perish as soon as hatched.   Also sprinkle the bed with coarse salt, which will not only destroy the grub but also act as manure to the roots.

In dull weather mildew may be found in a spring onion bed.   The only thing likely to do good is to dust the plants with sulphur, and a sprinkling of nitrate of soda would help the plant to grow out of it.

The best kinds to grow are : Sutton's Giant Blood-red Rocca, Sutton's Silver Globe, Sutton's Giant Zittau and Bedfordshire Champion (both good keepers), Ailsa Craig, Cranston's Excelsior, Trebons, James Keeping.

*RECIPES FOR COOKING*

## Onions à la Don Carlos

Take a Spanish onion, and cut a thick slice from the base, parboil it, then throw it into cold water, dry it, and scoop out the inside, and fill it with a mixture of pork sausage-meat and mashed potatoes in equal parts, mixed together with the yolk of an egg; have ready a small piece of boiled ham, and place under the onion; add a good brown gravy, and serve hot.

## Onion à l'Etoile

Take a large onion, core it with the column cutter, then cut it down within two inches of the bottom in divisions like an orange, so that it has somewhat the appearance of a star. Make a mince of beef or mutton, flavour it highly, and fill in the onion with it; tie the onion round in its original form until it is cooked, to prevent the mince falling out; and when it is cooked cut the string, and dish up with brown celery sauce round it.

## Onions Cornish Fashion

Boil some onions, chop them up, add half their bulk of mashed potatoes, and a fair proportion of butter, salt, pepper, and finely chopped parsley. Mix all well together, and put into a baking dish, and when it is quite hot serve immediately.

# PARSLEY

This is a biennial, a native of Sardinia, and was known in England prior to 1548.

Parsley is much more difficult to grow well than most people imagine, and good, deep, rich soil is needed, well trenched and manured. The best time for sowing it is at the end of February, and then in March, June, and July. The ground should be tolerably dry, and, as the seed is very small, it should be drilled shallowly in a fine seed bed, and the seed covered with three-quarters of an inch of fine mould, and then top-dress with soot as soon as the plants show through the surface.

Parsley will stand a moderate degree of frost, but sharp ones cut it down ; therefore it should be covered with litter during the winter.

For winter it is always desirable to place hand glasses over portions of it during the winter, or potting up bundles of the roots and place them into gentle warmth.

Soot is an excellent manure for it and preserves it from root-canker, the only disease which affects it. The plants do not seed till they are two years old. There are three kinds of parsley—the common plain-leaved, the curly-leaved, and the fern-leaved. The latter is a distinct foliage, and used for garnishing.

The best kinds to plant are Sutton's Imperial Curled, Sutton's Emerald Gem, and Fern Leaved.

## PARSNIPS

The parsnip is a very nutritious root. It contains three times more flesh-forming and strength-giving qualities than the potato, and only half the quantity of starch.

The parsnip is a hardy biennial plant, a native of England, and is almost frost-proof except in very damp or heavy soils.

The soil should be rich, deep, and loamy. An important point is that the soil should be *deep*, so that the root can strike down to its fullest extent, or otherwise the long, straight roots cannot grow properly.

For sowing, the ground should be deeply tilled in autumn and left rough during winter, and in March the ground should be levelled, and then trenched to a depth of about three feet. Plenty of rather rough littery manure should be worked in at the bottom of the trenches, but the surface soil must not contain anything of a rank or coarse description. Burnt earth and rubbish, leaf mould, and old rotten hotbed manure are the most suitable dressing for the surface. The seed should be sown about the middle of March unless the ground is wet and cold, as those circumstances would be fatal to it. The drills should be shallow and eighteen inches apart; cover lightly and finish with the rake. As soon as the plants are large enough, they must be thinned out to ten or twelve inches apart, or a little more if very large roots are required.

The ground should be kept very clean, and constantly looked over to remove all seedlings that may start afresh, and frequently hoed, and all weeds must be removed. The roots do not attain maturity till October, which is known by the decay of the leaves. If necessary, a *little* nitrate of soda sprinkled between the rows just before a shower greatly stimulates the crop.

Parsnips will keep in good condition if left in the ground and used as wanted, instead of storing them. They must, however, be all out of the ground before growth recommences in the spring, otherwise they will vegetate.

The best kinds to grow are Guernsey or Jersey Marrow, Elcombe's Improved, and Hollow Crown.

## *RECIPES FOR COOKING*

### Fried Parsnips

Boil the parsnips, after scraping them, in salt and water, drain, and dry them, and cut into lengthwise slices ; dip in melted butter, mix some salt and pepper into some flour, and fry.

### Parsnips à la Crème

Take two large parsnips, and boil till quite tender ; drain and cut them into finger lengths, cook them a few minutes longer in white sauce, then pour in a cupful of cream, and add two hard-boiled eggs ; cut into small pieces, and serve quickly.

## PEAS

The pea is a native of the South of Europe, but was not known in France till the middle of the sixteenth century. They were grown but never eaten till a gardener named Michaux first introduced them as a table vegetable, and was then brought into England in the reign of Henry VIII. It is one of our most important crops, and must be liberally treated to be successfully grown. A deep rich soil, well pulverised and well manured, should be chosen for the main crop in summer. For the early peas the ground does not require to be so rich.

From February on to June successional sowings should be made. In sowing them on light dry soils it is best to make deep trenches in which a heavy dressing of manure is put, with only a few inches of soil on the top to receive the seed. The distances between the rows must be regulated by the height of the varieties grown, the taller varieties five or six feet, those of ordinary height three or four feet apart. As soon as the seeds are sown, place pea-guards all over them, to prevent the birds eating the seeds, as the sparrows make great depredations on them. If pea-guards are not used, the only other way to protect them is to place a short stick at each end of the row, and then fasten from one to the other a single black thread at a distance of two or three inches above the ground ; but the galvanised pea-guards are the surest means, and they come in very usefully at other times, and last for years. Sticking the peas should commence as soon as the peas are three inches high.

The dwarf kinds can be grown without sticks, but even they are benefited, being kept from the ground. Before sticking them they should be hoed, and earth drawn round their stems.

When the summer sowings are made, if the weather is dry, the seed should be soaked in water for two or three hours previous, and the drills well watered.

Slugs are great enemies to peas ; therefore soot and lime should be scattered round the young plants on their first appearance, and greasy cabbage leaves placed near quicklime, thinly put over the ground where the peas are to be planted, and then forked over, is a good plan, and soot dusted over when they begin to appear. In dry weather they must be *well* supplied with water, and I consider Sutton's advice to open a shallow trench about a foot distant from the rows on the shady side, and pour the water into this, so as to fill the trench, is the best plan ; and if they are planted in rows, watering should be done between them.

Peas are generally eaten when not more than a quarter ripe, and should be used as soon as possible after gathering, as they injure by keeping more than any other kind of vegetable.

It is a good plan to top the leading shoots of the early crops as soon as they are in blossom, as it greatly hastens the setting and maturity of the fruit.

*Great* care must be taken when gathering the peas not to injure the haulm or stem, which causes the loss of many young plants. Johnson recommends the pods should be cut off with the scissors, as the plants produce one-fourth more than when roughly gathered from ; and if regularly gathered from, the longer they continue their production, as the later pods never attain maturity if the earlier ones are allowed to grow old before they are gathered.

The best kinds to grow are :—

American Wonder and Kentish Invicta for the earliest sowing ; Veitch's Ne Plus Ultra and Laxton's Fell Basket. In the second sowing and for the main crop, Veitch's Perfection, Champion of England, and Victoria Marrow.

The Duke of Edinburgh Pea grows very late, and keeps on growing, flowering, and podding continuously, affording an unbroken supply for eight or ten weeks at a stretch.

For those who have not much space the Exonian is very good : it grows very compactly, and to about three feet in height. Late peas ought to have a deep and fairly rich root run, and the soil so drawn up to the rows to form a basin, holding the copious waterings that must be given in August and early part of September.

## RECIPES FOR COOKING

### Croquettes of Peas

Boil some peas till tender in plenty of water, with a large bunch of mint and a tiny piece of soda. Drain very thoroughly, then beat to a pulp with a wooden spoon ; to every four tablespoonfuls of pulp add one ounce of butter, half a teaspoonful of pepper, ditto of salt, one tablespoonful of cooked bacon cut into very tiny shreds, the beaten yolk of one egg ; mix well together, and add sufficient breadcrumbs to make a panada ; roll into small balls the size of a nectarine, egg and breadcrumb them twice, and fry a nice golden brown ; pile in the centre of a dish, garnish with fried parsley, and surround with a purée of green peas made with cream.

### Peas à la Française

Put a quart of peas with two ounces of butter and one of flour into a stewpan, and mix over the fire till the butter is melted. Add a sprig of parsley, one or two small onions, a bunch of mint, a little salt and sugar, and a pint of stock. Put the lid on, and stew the peas for half an hour over a slow fire. Then drain them on a colander, remove the parsley and onion, add a little more butter, and serve.

## PEPERONI

This is a plant little known to England, and is much grown in Italy, and Naples in particular. It is a large glossy-looking fruit, very hot and sweet, and the skin resembles the tomato, and, like that plant, there are the red and yellow species. They will grow quite well in England, and should be grown in the same way as tomatoes, and I should advise anyone having a small greenhouse to try and grow them, as a variety of relishing dishes can be made from them.

They can be stuffed as tomatoes, and if dished up, the two colours alternately, they form a very pretty savoury.

## POTATO

The potato was first brought to England from Virginia by Sir Walter Raleigh, which, of course, is known by everybody. Loudon says it was formerly called the Potato of Virginia, and was thought at first very inferior to the Convolvulus Batatus, which was called the Spanish Potato, and the Jerusalem Artichoke, which was called the Potato of Canada.

Very few gardens give space enough to cultivate potatoes for the entire household use, but where space can be found, I should advise their cultivation.

The soil to ensure a full yield of potatoes should be moderately rich, yet the use of ordinary farmyard and stable manure to more than a very moderate extent has, though increasing the quantity of the crop up to a certain point, a very hurtful effect upon the quality of the tubers, rendering them soft and watery, and incapable of being kept for any length of time. Dwarf varieties will bear more manure uninjured than such as possess a strong habit of growth.

The best time for planting potatoes is in March,

nearer the end than the beginning. Plenty of space should be given between the rows (two feet for early varieties and two and a half to three feet for second early and main crop), and planted near the surface and earthing up deeply, which are most important, as the haulm makes strong growth and the tubers develop freely in the freshly worked soil, whilst the superfluous moisture is carried off by the deep furrows on each side of the row. The ground intended for potato cultivation should be deeply dug before Christmas with a heavy dressing of leaves buried in it, which should have been lying by for some time previous ; then the shallow drills should be made and the sets put in in rows of twelve to fifteen inches apart. Then cover in with a hoe, and as the growth appears above the surface the soil should be drawn up round them and the weeds hoed out. The final earthing up should not take place till the haulms have grown a good height, and then it is a good plan to sprinkle a little artificial manure between the rows.

Frost is a deadly enemy to potatoes, and the morning frosts in May are very inimical to them ; one hard frost will blacken and almost destroy them.

If light dry litter be placed a few inches over them it will be found a great safeguard to frosty weather. Sutton advises a few stout pegs on which mats can be laid as being most efficacious.

It is needless to say here that potatoes are liable to the potato disease, as it is only too well known the murrain or disease is a potato fungus, and is supposed to be the result of much wet and cold at the period of closing growth, when all bulbs and tubers require an extra degree of dryness and warmth ; the stem decays, and a smell of putrefaction is present, and the increased specific gravity of the potato amounts to one-third more than that of a healthy tuber ; when boiled the potato turns black.

Many gardeners slice the tubers into pieces containing two or three eyes, and some cut the tubers directly into halves, but the best results are obtained from planting them whole of medium size.

**Saving the Tubers.**—The medium sized tubers should form the most prolific roots for planting next year. Dry them by exposure to the sun for two or three days till they become slightly green, then store them in a dry place, laying them out thinly, and looking over them occasionally to remove any diseased ones.

The tubers should not be dug up till the tops are quite dead, and then a dry day should be chosen. Be careful in lifting the skin is not bruised. Rub off any sprouts that may grow till February, when the tubers should be set on end in layers in shallow boxes, and placed in a light cool shed where frost cannot penetrate ; the sprouts will then begin to grow strong, and when planted these should be half an inch long; but it is a good plan for amateurs to buy the tubers from Sutton or some other good seedsman to begin. When they have some experience in potato growing they can then save their tubers for seed.

**Potato Disease.**—A solution of sulphate of copper and lime, mixed and applied with a syringe before the disease has attacked the plants to any extent, will mitigate its further progress and ensure a healthy growth. The proportions are 22 lb. of the sulphate and 22 lb. of lime to 100 gallons of water to an acre of ground. The lime must be fresh and quick, and the purest sulphate of copper, and well mixed. Give one dressing before the disease appears at all on the foliage, about the first or second week of July, and a second one three weeks later.

Some of the best kinds to grow are—

Myatt's Ash Leaf Kidney—a fine old variety and early.

Early Rose—early and abundant.

Beauty of Hebrew—early and heavy cropper.

White Elephant—second enormous cropper.

Reading Giant—main crop very fine and very abundant cropper.

Schoolmaster—main crop very fine.

Rosette—late crop.

Magnum Bonum—main crop.

Scottish Beauty—late crop.

Sutton's Masterpiece.

Imperatum.

**Preserving Potatoes.**—Plunge the tubers before storing them away for ten hours in a 2 per cent. solution of commercial sulphuric acid and water—2 parts of acid to 100 parts of water. The acid penetrates the eyes to a depth of about one-fortieth of an inch, which serves to destroy their sprouting power. It does not have any appreciable effect on the skin of the potatoes. After remaining in the liquid ten hours, the tubers must be thoroughly dried before storing away. The same liquid may be used any number of times with equally good results.

This process has been adopted by the French Government for the preservation of the potatoes for the army.

## Pommes de Terre Farcies

Bake the potatoes, and, when nearly done, cut off a circular piece from the upper part, and scoop out a portion of the pulp, leaving about an inch of thickness under the peel. Then have ready some well-minced fricassee or forcemeat, butter the inside of the potato, and fill up the cavity with the mince heaped up in a point at the opening ; touch it over with raw yolk of egg, and put the potatoes again into the Dutch oven or brown them with a salamander. The skins should be rubbed with butter to render them crisp, or they are likely to become too hard to be peeled without breaking the potatoes ; but if not, a portion of it should be cut off.

## Potatoes à l'Italienne

Take enough mealy potatoes to make a good dish boiled dry ; two tablespoonfuls of cream, one tablespoonful of butter, salt, and pepper ; two eggs, yolks and whites beaten separately.

Whip up the potatoes while hot with a silver fork, which dries the potato from all superfluous moisture. When it is fine and mealy beat in the cream, the butter, salt, pepper, and whip up to a creamy heap before mixing in the whites, which should be first whipped stiff. Pile irregularly upon a buttered pie-dish ; brown quickly in the oven ; slip carefully on to a heated flat dish, and serve.

## Princess Potatoes

Boil one pound of potatoes, and steam them dry as possible, pass them through a wire sieve, and mix with them an ounce and a half of butter, one yolk of egg, and one ounce of grated Parmesan cheese, a pinch of salt, and a dust of cayenne pepper. Mix well together, and when cold roll the mixture into little rolls, using a little flour while doing so to prevent the mixture sticking. The rolls should be an inch and a half long and an inch in diameter. Place them on a buttered baking tin, brush over with a whole beaten-up egg, and fry them till a pale gold colour. When dished up pour a little warmed butter over them, and sprinkle over with finely chopped dried parsley.

## RADISHES

Radishes were introduced into England from China before 1584.

The Greeks esteemed radishes above all other roots, and it is recorded that in the oblations of garden fruits which they offered to Apollo in his temple at Delphos, radishes were presented in beaten gold, whilst other kinds were offered in lead and silver.

The way to obtain good radishes is to grow them quickly in rich soil and with enough moisture to keep the roots swelling steadily and freely.

The quicker a radish is grown the more delicious it is. The seeds should be sown in a rather light soil in an open sunny position, the soil dug deeply, burying plenty of half-rotted manure down in the bottom of the trenches, where it should lie for two or three weeks longer for the weather to act upon and to render the surface fine. It should be raked to a fine even surface, adding road scrapings or leaf mould, with a sprinkling of soot. The seed should be sown broadcast, but not too thickly ; rake it in lightly, and where the soil is light, make the surface firm with the back of the spade. It should be kept moist but not wet. During the summer a sowing should be made about every ten days to produce a regular supply.

From October to February sow in frames ; after that sowings may be planted in the open on warm borders, with a little light litter or fern over, or matting supported by sticks.

When the seedlings are well up they should be thinned, and in dry weather should be watered regularly every night. The seedlings are in general up in less than a week, and in six weeks they should be ready to take.

When the plants are in frames the glasses should be closed of an afternoon, and, whenever the earth appears dry, light watering must be given at noon.

There are several varieties of the radish—spring varieties, autumn and winter varieties, and turnip-rooted and long radishes.

The best all round kinds are the French Breakfast Radish, Sutton's Early Crimson, Sutton's Gem, Sutton's Black Summer (turnip), Sutton's Long Rose, and Black Spanish for winter varieties.

## *RECIPE*

### Radish Sandwiches

Slice turnip radishes very thinly, and lay them on bread-and-butter, with a little shredded lettuce dipped in oil and vinegar ; add pepper and salt, and place between slices of buttered bread in the usual manner.

## SAGE

This herb thrives best in a light and well-drained and sandy loam, and is usually raised from cuttings formed of the young side shoots taken in April or May, inserted under a hand-light, and kept slightly moist and lightly shaded until started. A fresh stock ought to be raised every three or four years, as after that time the old plants get worn out and straggling. If the plants are cut down too closely they die out.

## SALSIFY AND SCORZONERA

To grow them the ground should be prepared in autumn in the same way as for parsnips, and the seed should be sown in drills fifteen inches apart, and the plants thinned out afterwards to eight or ten inches apart in the row. Hoeing should go on perpetually in the summer, and in winter salsify should be taken up and stored the same as for carrots.

Scorzonera is similar to salsify, and should be treated in a similar manner.

Unless a garden is fairly large I would not advise either of these being grown, as they are not to everybody's liking.

### *RECIPES*

### To Dress Salsify

Scrape the roots, and lay them in cold water for fifteen minutes. Boil whole till tender, drain, and when cooked mash them to a paste, taking out all the fibres ; moisten with a little milk. Add a tablespoonful of butter, and an egg and a half for every cupful of salsify. Beat the eggs light, make into round cakes, dredge with flour, and fry brown.

## Scorzonera in Brown Sauce

Wash and scrape the skin off gently from the roots, and cut them into lengths of three or four inches ; put them into boiling water with a little salt butter and the juice of a lemon ; boil for an hour, then drain, and serve with a rich brown sauce over them.

# SAVORY

This consists of two kinds, the summer, which is an annual, and should be sown in April and afterwards thinned out ; and the winter savory, which is increased by division of the roots in April.

# SEA KALE

Sea kale is one of the most delicious vegetables grown for culinary purposes, and those who possess a small vegetable garden should grow it, for it is not at all difficult, and is generally expensive to buy.

Sea kale plants are propagated in two ways, one by sowing seed and the other by means of root cuttings ; but for amateurs I advise to raise from cuttings, as it takes two years to raise plants from seeds. In the first place, I should advise buying a few full-grown *crowns* and planting them, and then from them take cuttings or seeds for another season. March is the best time to plant the roots. It will succeed in almost any soil, provided it is deeply dug and *well* manured. The ground should be trenched two feet deep, and as much vegetable refuse and road sand should be worked in within one foot of the surface.

When the roots are received early in March cut off the crown just below the junction of the leaves with it, or else a truss of bloom will grow from each crown and will interfere with the expected crop.

When the crown is cut off, two and sometimes more shoots spring from it, which should be thinned down to one, selecting the strongest. Plant the roots (three) anglewise six inches apart and two feet six inches between each clump. Cover the heads with two inches of soil. Keep the ground from weeds and continually stirred with a Dutch hoe.

In summer it should have plenty of water, liquid manure, and mulchings of rich stuff.

As the leaves decay in autumn they must be taken away and the ground kept clean and manured.

The sea kale pots should be uncovered every now and then to see how it is growing, and some days left a little open for a few minutes to let in air to them ; but darkness is required, so that the kale may grow perfectly blanched.

**Forcing.**—The crowns should be lifted about the middle of October and then placed into the earth and given a month's rest before planting them again ; then in the middle of November have a box or boxes about two feet square and eighteen inches deep, and put in the bottom six inches of soil and place the box in a temperature of 60°, and keep it there till the sea kale is ready, which it ought to be in a month's time. At the beginning of the year sea kale pots should be placed over the crowns.

When preparing the roots for forcing, save the best pieces of the trimmings to be used as sets in the spring, and thus keep up the stock. All pieces the thickness of a cigarette should be thrown by themselves and stored in sand in a shed till the spring, when they should be planted out, and the season after they will be again fit to use.

When sea kale is forced in the open ground it should be covered with sea kale pots, and surrounded with fermenting manure, or buried in a couple of feet of half decayed leaves, and these again with hot manure. Salt is a very good fertiliser to use.

Before placing the pots on, each crown should be dusted over with quicklime to destroy all worms and slugs.

The heat had better be too low than too high ; a few coal ashes are good to use in the coldest weather, just enough to cover the crowns, which prevents the slugs eating into the crowns when the kale is cut.

The sea kale pots, when not in use, can be utilised for planting geraniums and other summer plants in, and stood in front of a greenhouse, or down garden walks, or at entrances to the house.

Sea kale should be cut near enough to the crown for the head to remain intact. The most usual method is to cut the whole crown off bodily as soon as ready.

## Sea Kale au Jus

Boil the sea kale in salted water, drain it, and put it into a saucepan with as much good gravy as will cover it, and stew till tender. Lay the sea kale into the dish it is to be served in, and boil up the gravy ; thicken with a little butter and flour, squeeze in the juice of half a lemon, and pour it over.

# SHALLOTS

February is a good time to grow shallots ; if the soil is stiff it should be heavily dressed with fresh manure in autumn, but if light, well rotted manure should be dug in at planting time. Shallots like a firm soil ; a little lime worked in does wonders in shallot growing. The beds should be four feet in width, and raised one foot or eighteen inches above the paths. The bulbs should be planted shallow, nine inches apart and one foot between the rows. They must not be pressed *too* far into the ground though, as the roots go pretty deep into the soil. Birds and worms frequently lift the bulbs from the ground ; therefore it is well to go over the bed occasionally and replace those removed. Frequent use of the Dutch hoe is necessary to keep the soil free, and weak liquid manure should be occasionally given during the growing season. The best bulbs to grow are those of middle size. The best kinds to grow are Yorkshire Hero, Jersey, and Sutton's Giant.

# SORREL

Sorrel thrives best in any rich, light, garden soil, and is propagated by seed and by parting the roots. The seed should be sown in March and April, and the roots parted in September and October. To sow the seed make drills six inches apart and a quarter of an inch deep, and when the seedlings are two or three inches high, they should be thinned to three or four inches apart. The bed must be kept free from weeds and watered occasionally in very dry weather.

In September the entire crop should be transferred to fresh ground, allowing eighteen inches between the plants, or part may be withdrawn.

To divide the roots, take and set them at once where they are to remain, in rows eighteen inches apart. In summer the stalks must be cut down to encourage the production of the leaves.

The French sorrel is the best sort to grow.

*RECIPES FOR COOKING*

## Sorrel à la Française

Take some sorrel, pick off the stems, and wash the leaves in several waters, then put them into a stewpan with a pint of boiling water in which some salt has been dissolved, and let boil till tender ; then drain it and rub it through a wire sieve. Put into a saucepan with plenty of butter, a dessertspoonful of flour, and a little pepper. Stir over fire till it boils. Add three tablespoonfuls of cream.

## Sorrel Fried in Batter

Make a small quantity of frying batter, and take the *middle* leaves of the sorrel ; wash them in two or three waters, drain them, and trim them. Dip them into the batter, and fry in hot fat till crisp and brown. Serve hot, garnished with fried parsley.

## SPINACH

Spinach is very easy to cultivate ; it only requires to sow the seed broadcast or in drills in moderately rich good ground, loam or clay, and when up thin it out to a few inches apart. If the plants run to seed soon after it comes up, as it often does, the reason is the ground is too light, poor, or *dry*, and weather too hot, and in this case the ground requires to be made deeper and richer. There are four kinds of spinach, the smooth seeded or summer spinach and the prickly seeded or winter spinach, Mountain spinach and New Zealand spinach. The first should be sown in drills a foot apart and an inch deep every three or four weeks from the middle of February till August, and when well up to be thinned out to six inches apart ; keep it moist at the root.

The winter spinach should be sown in August and up to the first week in September. The soil should be firm and moderately rich, and the plants thinned out to nine inches apart.

Watering the plants with a solution of nitrate of soda, the strength of an ounce to a gallon of water, produces large leaves in abundance. Soot is good to put to the roots also.

New Zealand spinach is sown in heat in March

or April, and the plants turned out in June into very rich soil in a sunny situation, allowing plenty of room.

Mountain spinach, also called Arach, is a hardy annual, and grows very large ; it should be allowed a distance of about eighteen inches, and in rows two feet apart.

There is a variety of the summer spinach called Flanders spinach, with large thick leaves, and another called lettuce-leaved, the leaves of which are still larger and very dark green in colour.

For small gardens the perpetual spinach, or spinach beet, is most useful, as it requires so little trouble ; it is a totally distinct spinach from the other kinds.

The summer crop, when gathered, should be pulled up by the root unless it is *perpetual*, but the winter crop should only have the outer leaves gathered, and it will then continue producing fresh leaves for many months.

The best kinds to grow are Perpetual, or Spinach Beet ; round for summer use, prickly for winter use.

## RECIPE FOR COOKING

### Spinach à la Française

Boil the spinach with scarcely any water to it (but it must be boiling) and just a piece of soda the size of a pea, and a little salt ; boil for ten minutes, press down with a spoon, drain and squeeze, and throw it into cold water to preserve the green colour. Drain, and pass it through a wire sieve, and then return it to the saucepan with a little salt, a piece of fresh butter, an eggspoonful of castor sugar, the squeeze of a lemon, and a gill of cream. Heat up in the saucepan after stirring it well ; press it into a plain mould and turn it out, or else, after turning out, cut it into the shape of diamonds and arrange it *en couronne*, like cutlets.

## TARRAGON

This herb is propagated annually by means of cuttings or division in the spring, say about March or April, and the pieces planted eight inches apart in rows, covering them rather more than two inches deep, or cuttings may be taken in. July and rooted under a hand glass, keeping them moist and shaded. For winter use the plants should be cut over in August and lifted, and potted in September, keeping them in a cool frame for a time, then removing to a greenhouse shelf, where they will continue growing all the winter.

## THYME

There are three kinds of variety of thyme, the broad leaved, the narrow leaved, and the variegated, and grows best in eight-inch soil, and should be divided occasionally to keep the plants vigorous. Seeds should be sown thinly in pans of sand and rich soil in March under glass and planted out in May, though it is best to propagate by cuttings and to buy the roots.

## LEMON THYME

This herb is of much larger growth, and possesses a most delightful odour and flavour, and is most generally preferred to the other. They are both increased by division of the tufted shoots, and the rooted bits planted out singly in March or April ; layered shoots also strike readily.

## TOMATO

Tomatoes can be raised from seed in any ordinary hotbed or shelf in a greenhouse. The hotbed should be about three feet high at back and two and a half feet at front, and the seeds should be sown broadcast. They come very strongly treated in that way. As soon as the plants come up, whenever the weather is mild they may be sown in pots and the plants pricked off when large enough. After potting off warm covering will be necessary at night. The chief requirements are some good rich soil composed of loam, leaf mould, and manure, two parts of the former and one of the latter. Plant from twelve to eighteen inches apart ; air should be freely given in all stages. When the plants are well set with fruit, liquid manure should be given two or three times a week.

For planting tomatoes in the open, seeds should be sown not later than March and placed in a greenhouse or frame, and when the plants are up and thinned, and the time for planting arrives, they can be planted out not later than June against a sunny wall. The plants ought to be turned out with the ball entire, only just loosening a few of the roots to enable them to get hold of the new ground quickly.

When the plants are making strong growth, and if they stand at all closely together, it is well to shorten the lower leaves when the plants have commenced fruiting, as it is desirable to let in air and sunshine ; half a leaf will be better to cut away than cutting away altogether.

If dark spots appear on the tomato leaves in the greenhouse, the plant will be affected with the tomato disease, which are fungus growths brought on from want of ventilation, or with shutting up the house closely at night, with too much moisture about. They will require freer ventilation, and the under sides of the leaves dusted with sulphur and the atmosphere kept dry, especially towards evening. Cut away and burn all the worst affected leaves.

In excessively hot weather, where the tomatoes are under glass, they require an immense amount of water, and the difficulty is to give them enough ; they should never be watered less than twice a day.

It is a good sign that the plants are healthy and doing well when some of the older leaves curl up.

If the plants are very luxuriant the secondary shoots should be stopped.

Training the plants may commence as soon as the branches are a foot long, and continued throughout their growth. Where there is no wall or palings to train them on, they may be trained with stakes as espaliers. The plants should never be topped whilst there is head room for them, as early topping curtails their productiveness greatly.

If tomatoes do not colour, and it is getting too late for them to do so, if they are of a usable size they should be gathered and placed in a sunny window, or on the ledge of a warm house, or on shelves in the kitchen they will ripen and colour well, or they may be cut in bunches with a portion of the stem attached and hung up wherever there is warmth.

Tomatoes are liable to be affected with the disease known as *Cladosporium fulvum*, which is generally the result of a too close warm and moist atmosphere. The worst affected leaves should be cut off and burnt, and if the plants are at all crowded, shorten back all the lower leaves considerably to admit light and air. The house should be ventilated night and day, and on wet cold dull days a gentle fire heat should be used, and only just enough water to keep the plants fresh and growing.

A little nitrate of soda dissolved in the water once or twice a week will help.

The best varieties for the open ground— Laxton's Open-air, Mikado, Carter's Greengage, Sutton's Golden Nugget, Golden Eagle, Sutton's Dessert.

Against walls —Perfection, Golden Queen, Horsford's Prelude, Conference.

The latter two are deliciously sweet, and two of the best setters ever raised.

Hathaway's Excelsior is a capital tomato, grows both under glass and in the open.

The very small fruited varieties are Red and Yellow Cherry, and Sutton's Cluster Red Currant.

*RECIPE FOR COOKING*

## Tomatoes Farcies

Take six tomatoes, cut a circle off the top of each, scoop out the insides, mince some shallots very fine, and toss them in butter with some finely minced parsley and a little grated Parmesan cheese ; add the tomato pulp to this, and then fill the tomatoes; put on the tops again, and bake in a moderate oven.

There are so many delicious recipes for cooking tomatoes that it is difficult to select any in particular here, but I have a *very* large collection of many ways of cooking them, some of which will be found in my ' Savouries à la Mode ' and ' Dressed Vegetables à la Mode,' from which tomato lovers can choose for themselves, and a great many in my *manuscript* collection of vegetable cookery, foreign and English, any of which I will send to my readers if they like to send a shilling, which will be devoted to help young gentlewomen in perfecting themselves to endeavour to gain their own livelihood.

# TURNIPS

Turnips are not very easy to grow, especially during a dry season. The principal things are deep digging and liberal manuring, keeping the soil moist, and encouraging a rapid growth by means of frequent top dressings of wood ashes, soot, salt, and lime. The seeds should be sown in March and April, and on a plot of ground that has been deeply dug and heavily manured, and occupying a lightly shaded position ; but although deep and rich below, the bed must be quite sweet on the surface.

The seed is generally sown broadcast, and the

seedlings when up thinned out to about a foot apart.

The ground should be frequently hoed after the plants are up, so as to keep down weeds and air the soil to check evaporation. Burnt earth applied as a top dressing three or four times during growth has a wonderful effect upon the growth of the plants.

A second sowing is generally made at the end of May and a third for the main crop towards the end of June.

For early sowing, Sutton's Early Snowball, Early Milan, and Carter's Jersey Lily ; and for main crops, Veitch's Red Globe, Orange Jelly, White Stone, Red American Stone, and Chirk Castle Black Stone. For autumn and early winter supply, Orange Jelly and Chirk Castle Blackstone are best. The skin of the latter is black, but the flesh is white as snow, and possesses a finer and more delicate flavour than any other, and is perfectly hardy.

There are also many kinds of turnips of French origin, such as the Long Forcing Paris, which resembles a large oval radish, the Long White Means, or Cowhorn Turnip, with long crooked roots like a deformed carrot, and the Yellow Finland.

Turnips are very liable to the turnip fly pest, which makes great havoc among the young plants directly they appear above ground in dry weather. The principal safeguards are keeping the soil moist and frequent dustings of lime given when the foliage is wet with dew or rain. They are also liable to a disease called finger-and-toe, which is a fungus, and the only thing to do is to give a little gas lime.

*RECIPE FOR COOKING*

## Curried Turnips

Take some cooked turnips, fry them in butter with a little onion and a tablespoonful of curry powder, a teaspoonful of sugar, one of desiccated cocoanut, a grate of nutmeg, and half a teacupful of milk. Let all stew slowly for an hour. Serve with rice.

## VEGETABLE MARROWS

Vegetable marrows should be sown in April under glass. A temperature of 55° or 60° will soon cause the seed to vegetate. As soon as the plants form a rough leaf, they must be potted off singly in six-inch pots. Return them to a frame till re-established. The plants may then be gradually hardened off by opening the frames in the daytime, and their final planting out should take place at the end of May. It is of no use putting them out earlier, as they are so easily hurt by frost.

When they have made a good start, the shoots must be stopped to make each throw out from six to eight leading stems, which can be trailed in different directions. As soon as fruit is set extra nutriment can be supplied to any extent by means of frequent applications of liquid manure, and, in dry weather, this assists the plants materially.

Marrows should be cut, when cut, young, as the flavour is not only more delicate, but the plants will yield several times as many as when allowed to grow to a great size.

The best kinds to grow are Sutton's Long White, Sutton's Long Cream, Improved Green Bush, and Improved Custard.

*RECIPE FOR COOKING*

## Vegetable Marrows Fried

Take a couple of marrows, pare, seed, and quarter them, salt and pepper them, then let them marinade in a wineglass of vinegar and double the quantity of salad oil, and let them remain in it for half an hour. Then drain them, cover them with poivrade sauce, and stew till tender, and when ready sprinkle some grated Parmesan over, and serve quickly and hot.

# WATERCRESS

Few people think of growing watercress. They are quite contented to buy it; but it is so easy of culture, and its medicinal properties are so great, that I think that those who have gardens would find it most useful. In the days of our forefathers it used to be grown, as it was an important ingredient in 'spring tea.' The 'Gardener' writes: 'An individual named Bradbury, during the seventeenth century, was the first to cultivate it in this country (it hailed from Holland in the first instance). This man made use of some natural watercourses running through his market-garden for the purpose. He simply sowed the seeds in a ditch, and regulated the height of the water by means of dams of earth at regular intervals.' Where there are running streams it is a very easy matter, but it is not necessary to have a stream to grow them in, as they are easily cultivated in borders, and this must be done in September in a moist and shady border. The earth must be dug fine, and a slight trench drawn with the hoe, and fill this with water till it becomes mud.

Cover it about an inch deep with drift sand, and then stick in the slips about six inches apart, watering them till established. The sand keeps the plants clean. They will be ready for gathering in a very few weeks, and must be cut and not picked. For planting in water see Johnson's 'Gardener's Dictionary.'

## USEFUL HINTS

### ARTIFICIAL MANURE

Ichtheinic guano or Clay's Fertiliser are excellent manures.

Decayed leaves, with weeds or other green vegetable matter added to them in a heap, together with some kitchen refuse, make a good manure. Cabbage, broccoli, and turnip leaves benefit any kind of soil. The leaves of the Brassica family emit a bad smell when decaying, but this can be remedied by covering them in the earth.

### CROSSING VEGETABLES

The form and arrangement of the pistil and stamens vary much in different plants, and the signs of the former being in a fit state for inoculation are so different that only very general directions can be given in such a small treatise as this.

To insure a perfect cross the anthers should be removed with a pair of small pliers before they begin to shed their pollen. When the stigma arrives at the proper stage, which is known by various signs, the pollen from the plant selected for the male parent must be applied to it either by applying the bloom itself or by means of a soft camel's-hair pencil. Cucumbers and marrows are very easily dealt with, all that is necessary being to apply pollen from the male bloom to that of the female at any time while expanded.

It is best done early on a light day, when the sun is shining and the air is dry.

## DRYING OF HERBS

Herbs should be gathered on a fine day when they are at their full growth and before they toughen.

They should be cleansed from all grit and dust and dried in the oven as quickly as possible ; then pick the leaves and pound them to powder, each sort separately, and then put them into tightly corked glass pots.

Chervil fennel and parsley may be dried from May till July.

Mint-thyme and marjoram in June and July.

Lemon-thyme, summer and winter savoury, and tarragon in July and August.

Sage in August and September.

It is well to have one pot of mixed fresh herbs, such as two ounces of parsley, chervil, and mar-joram ; one ounce of basil thyme, lemon thyme, and savoury ; and half an ounce of tarragon.

## GRASS SEED TO SOW

Rake the lawn thoroughly first ; tear the ground up with the rake, and sprinkle fine earth all over very evenly ; put earth in a sieve to sprinkle it over, then sow the seeds and roll well ; roll about three times lengthways and crosswise ; when the seeds show about an inch above ground, the ground should be rolled regularly.

## HOTBED (TO MAKE A)

Stable manure and equal quantities of tree leaves make the best and most lasting hotbeds. The manure should be well shaken over and thrown into a heap to get warm. If there are no leaves to mix, the heap should be turned over, the outsides thrown into the middle, and the dry and damp spots well mixed. This is most necessary to give a lasting heat. A hotbed should be a foot wider on all sides than the frame ; the size and height must depend on what it is wanted for. The bed must be built up in layers, and after each layer pressure must be brought to bear on it as it is placed on the surface and well beaten down with a spade. The frame and lights should be put on immediately it is finished, and the bulb of a thermometer placed just inside the manure in the frame. The fresh material should be turned over twice or thrice for two weeks, and then the rank steam will be thrown away and the violent heat subsided. Nothing should be placed in the frame till the fiercest heat has passed off. It is fit for use when the heat does not exceed 80° or 85°.

## LETHORION FUMIGATOR, AND HOW TO USE IT

This is a first-rate fumigator, in the shape of cones, and all that is necessary is to remove the outside wrapper before burning the cone and see that the house is thoroughly secured before lighting the cone (thoroughly air-tight). A calm evening should be selected for the operation, and the cone must be well-lighted all round the top before leaving it in a house, when it will smoulder about half-way down in from ten to twenty minutes before emitting the vapour, which will continue until the contents of the small bottle inside the cone is exhausted.

In calculating the number of cones required for large houses a reduction of 30 per cent. in cone power, commencing from 3,000 cubic feet and upwards, will be found of sufficient strength for ordinary aphis. Thus, too, No. 3 cones for a 3,000 ft. house will suffice, but much depends upon the security of the house, as the vapour has great affinity for the outside air.


## LOAM

Is earth easily worked at any season, and being sufficiently retentive, yet not so much so as to hold water.

Maiden loam is the fat earth forming the top spit of a pasture ground, and that with a yellowish-brown is the best.

Sandy loams are the easiest to work and yield the earliest produce.

Chalky loams are early and fertile if there is not too much chalk.

Clayey loams are bad to work either in wet or dry weather, being wet and sticky in one case and dry and cracking in the other.

## NETTLES, TO DESTROY

Keep cutting them down, and they will not stand this for long ; fork out the roots if possible, but care must be taken not to injure the roots of any of the shrubs near. As soon as they appear above ground, cut them over just under the surface of the ground with a hoe, and every time they sprout, hoe them down again, and they will soon disappear.

## PROTECTING VEGETABLES FROM FROST

Cauliflowers showed the foliage bent down over their heads until nearly fit to cut. Then lift the plants and place them under cover out of reach of frost, or lay them in closely together out of doors, where mats and litter can be easily applied.

Celery keeps safely if *well* earthed up, and a covering of litter placed over the tops of the hedges keeps the frost from penetrating, and lifting is much easier done.

Carrots should be covered up with litter as soon as frost begins to penetrate.

Turnips will stand a good amount of frost but not if exposed to freezing right through, and it is a good plan to pull up the roots in December and lay them in trenches, covering the roots right over with soil but leaving the tops exposed.

## SEEDS, TO SOW

Seed-sowing is the chief and most general way of propagating plants. It is often more convenient to raise them by means of cuttings or layers, but vegetables especially are better increased by this means than in any other way.

A moderately free or light, porous, and yet fairly rich soil is most necessary for successful germination. The primary root of any seedling is very delicate, and the future vigour of the plant depends upon its strength and healthiness. If it strikes down deeply and easily into a free, rich, yet sweet moist soil bed, the young plant will flourish, but if checked in any form it will become unhealthy. The soil for seed-sowing must be mellow and fairly light for the young roots to penetrate it, *porous* to admit air to the feeding points, and rich, yet sweet to give enough nourishment to keep the seedling going.

A moderate degree of moisture and a suitable temperature are necessary conditions.

## SPARROWS

These are very destructive in most gardens, and if a few oats are scattered on a piece of ground and where limed twigs are placed they can be caught ; but the most humane way is to stretch black cotton backwards and forwards just above or at the side of any special plants, for then the birds catch their legs against the thread, which frightens them.

## INSECT PESTS

Rich earth is often the cause of wireworms, and grubs, and common slugs, who revel in good soil. The best plan for the destruction of the wireworms and grubs is to water with soap-suds occasionally and to carefully pick out all that can be seen when cultivating the ground. The same course should be taken to kill slugs, but in addition quicklime should be not only well mixed with the soil when the bed is made, but sprinklings should be given late at night, so as to catch them when they are busy at work.

It is a good plan to dress the ground intended for vegetable marrows with gas lime in the previous autumn, which kills the pests in the soil, besides partly manuring the ground.

## ANT TRAP

Soap a sponge in water and wring it nearly dry, then sprinkle it with sugar and lay it on a plate in the haunts of ants. When full, plunge it into boiling water.

Boiling water poured over them will kill all it touches. Pieces of raw meat, or bones from which the meat has been cut, have a wonderful attraction for them: these should be thrown into scalding water. A strong solution of Sunlight soap, six ounces or eight ounces to the gallon, will destroy any insects.

## ASPARAGUS BEETLE

This is a very pretty little beetle, but it does a great deal of damage, as it injures the foliage and checks the growth of the stems.

Their larvæ do the most harm, and should be hand-picked early in the season. The best exterminator is to sprinkle soot over the foliage when damp, mixed with salt ; the mixture should be 20 of the soot to 1 of the salt.

## BEAN FLY

Is an insect which attacks beans, and which is very destructive to the young shoots and foliage. The best means of ridding the beans of them is to remove the tops of the infested shoots and wash the plants well with soapy water.

## BLACK FLEA

Attacks turnips ; then cover the surface with gas lime two or three mornings after the turnip seed has been sown.

Blight is the popular name for any withering of plants such as occasioned by violent cold winds in early spring ; the ravages of the hawthorn cater-pillars are also spoken of ; but what it really is is not yet understood.

## CABBAGE FLY, OR CATERPILLAR

These are the common white caterpillars which hover over the cabbages in the spring and summer. They lay their yellow eggs under the surface of the leaves, and from these issue the green caterpillars. They are green in front and yellow behind at first, then they get hairy and dotted over with black.

Hand-picking is the best method of cleansing the plants from them. Soap-suds is a capital remedy, and if an ounce of paraffin oil is mixed with a gallon of soap-suds with which to drench the cabbages when quite young and before the caterpillars appear.

## THE CARROT GRUB

Is a dangerous enemy, and a dressing of soot and lime should be liberally supplied. A good watering with soap-suds all over the soil and round the stems of the plants will keep these pests at bay.

## CELERY FLY

Hand-picking is generally regarded as the safest remedy, and if every leaf is removed and burnt directly, the mark of the insect can be perceived from the first. It cannot increase to any extent. Syringing the plants vigorously with a mixture of water and paraffin, at the rate of six gallons of the former to a wineglassful of the latter, is an excellent preventive. The oil must be kept well mixed with the water by the frequent use of the syringe in the pail. The operation, to be effectual, must be performed only while the sun is shining, when alone the flies emerge from the feeding-ground in the leaves and proceed at once to lay a fresh batch of eggs. Another good plan is to dust with lime or soot to prevent the grub coming ; but when once the grub appears, it should be crushed by pinching the leaf. This pest causes what looks like blisters on the leaf.

# GREEN FLY (TO DESTROY, ANOTHER WAY)

Boil a quarter of a pound of quassia chips in a quart of water for half an hour, then stir in two ounces of soft soap and add enough cold water to make a gallon. Dip the affected plants overhead in this mixture, which will kill every insect directly it touches them, and five minutes afterwards syringe the plants well with clear soft water.

## MILDEW

Use the same wash as for green fly, with a handful of sulphur mixed in. Syringe in the afternoon and shut up close, and next morning put about an inch of rich soil on the surface. Repeat the syringing several times, but use the syringe gently.

## PEA BUG, OR WEEVIL

This pest attacks the peas at the time of flowering, or whilst the pods are setting.

## THE POTATO BUG

This lives upon the foliage of potatoes, and is said to be the cause of the potato disease.

## RED SPIDER IN A CUCUMBER HOUSE

Wash the plant with soft soap and water, and limewash the walls and mix in a large handful of sulphur with the lime. The red spider is a yellow or pale red colour, and congregates thickly on the underside of the leaves, and protects itself by means of a fine cobweb-like covering. They are caused by an over-dry atmosphere, and, if they appear early in the season, denote neglect of watering.

## SLUGS

One of the best ways of ridding plants of slugs is to get a little malt after it has been used for brewing. Place it on a piece of slate where the pests frequent, and they will eat it so greedily that they will swell to a great size and stiffen so much that they will die on the spot.

## WIREWORMS

These are most destructive, and once they get hold of any soil there is nothing but to dig it and pick them out. To prevent their attack upon a crop, mix a little spirit of tar on a larger quantity of gas lime with the soil.

To trap them, bury potatoes in the soil near the crop, and thrust a piece of stick through it as a handle to take them up by.

Salt and soot are the best things to get rid of these pests. If very badly affected, clear the ground, give in the autumn a heavy dressing of salt and soot, and let it be a month ; then throw it up in rough ridges for the winter, and there will be no wireworms the next year.

Or a smaller quantity of salt with a fair sprinkling of soot may be scattered among the plants, allowed to lie awhile, and then lightly forked in.   Much salt has the effect of rendering damp soil pasty and rotten, and soot in quantity will perish the texture of light ground, and make it like so much dust or rubbish, and act accordingly.

A dressing of gas lime spread over the ground and dug in is a good thing.

When salt is applied it should be in showery weather, and on a fine day stir in the surface of the soil.

Wireworms are particularly fond of pansies, and it is said to be a good plan to grow round the beds with an edging of daisies, as they will devour their roots with avidity.


## GARDENING  MEASURES

Half-sieve contains three imperial gallons and a half.

Sieve, seven imperial gallons.

Bushel sieve, ten and a half imperial gallons.

Bushel.—A bushel of cleaned and washed potatoes, weight fifty-six pounds ; but if not cleaned, sixty pounds should be allowed.

Brown's Fumigator is the best to have.

# INDEX